PRENTICE HALL SCIENCE

MOTION, FORCES, AND ENERGY

Anthea Maton
Former NSTA National Coordinator
Project Scope, Sequence, Coordination
Washington, DC

Jean Hopkins
Science Instructor and Department Chairperson
John H. Wood Middle School
San Antonio, Texas

Susan Johnson
Professor of Biology
Ball State University
Muncie, Indiana

David LaHart
Senior Instructor
Florida Solar Energy Center
Cape Canaveral, Florida

Maryanna Quon Warner
Science Instructor
Del Dios Middle School
Escondido, California

Jill D. Wright
Professor of Science Education
Director of International Field Programs
University of Pittsburgh
Pittsburgh, Pennsylvania

Prentice Hall
Englewood Cliffs, New Jersey
Needham, Massachusetts

Prentice Hall Science
Motion, Forces, and Energy

Student Text and Annotated Teacher's Edition
Laboratory Manual
Teacher's Resource Package
Teacher's Desk Reference
Computer Test Bank
Teaching Transparencies
Product Testing Activities
Computer Courseware
Video and Interactive Video

The illustration on the cover, rendered by David Schleinkofer, shows a soccer ball after it has been struck. It depicts motion, forces, and energy.

Credits begin on page 160.

SECOND EDITION

ISBN 0-13-402041-3

8 9 10   97 96 95

Prentice Hall
A Division of Simon & Schuster
Englewood Cliffs, New Jersey 07632

STAFF CREDITS

Editorial:	Harry Bakalian, Pamela E. Hirschfeld, Maureen Grassi, Robert P. Letendre, Elisa Mui Eiger, Lorraine Smith-Phelan, Christine A. Caputo
Design:	AnnMarie Roselli, Carmela Pereira, Susan Walrath, Leslie Osher, Art Soares
Production:	Suse F. Bell, Joan McCulley, Elizabeth Torjussen, Christina Burghard
Photo Research:	Libby Forsyth, Emily Rose, Martha Conway
Publishing Technology:	Andrew Grey Bommarito, Deborah Jones, Monduane Harris, Michael Colucci, Gregory Myers, Cleasta Wilburn
Marketing:	Andrew Socha, Victoria Willows
Pre-Press Production:	Laura Sanderson, Kathryn Dix, Denise Herckenrath
Manufacturing:	Rhett Conklin, Gertrude Szyferblatt

Consultants

Kathy French	National Science Consultant
Jeannie Dennard	National Science Consultant
Brenda Underwood	National Science Consultant
Janelle Conarton	National Science Consultant

CONTENTS
MOTION, FORCES, and ENERGY

CHAPTER 1   **What Is Motion?** 10
   1–1 Frames of Reference **12**
   1–2 Measuring Motion **14**
   1–3 Changes in Velocity **21**
   1–4 Momentum **25**

CHAPTER 2   **The Nature of Forces** **34**
   2–1 What Is Force? **36**
   2–2 Friction: A Force Opposing Motion **38**
   2–3 Newton's Laws of Motion **41**
   2–4 Gravity **47**

CHAPTER 3   **Forces in Fluids** **60**
   3–1 Fluid Pressure **62**
   3–2 Hydraulic Devices **65**
   3–3 Pressure and Gravity **68**
   3–4 Buoyancy **70**
   3–5 Fluids in Motion **74**

CHAPTER 4   **Work, Power, and Simple Machines** **82**
   4–1 What It Means to Do Work **84**
   4–2 Power **86**
   4–3 Machines **89**
   4–4 Simple and Compound Machines **92**

CHAPTER 5   **Energy: Forms and Changes** **106**
   5–1 Nature of Energy **108**
   5–2 Kinetic and Potential Energy **111**
   5–3 Energy Conversions **116**
   5–4 Conservation of Energy **119**
   5–5 Physics and Energy **121**

SCIENCE GAZETTE

Guion Bluford **130**
Robots **132**
Hypersonic Planes **134**

Activity Bank/Reference Section

For Further Reading   136
Activity Bank   137
Appendix A: The Metric System   153
Appendix B: Laboratory Safety: Rules and Symbols   154
Appendix C: Science Safety Rules   155
Glossary   157
Index   159

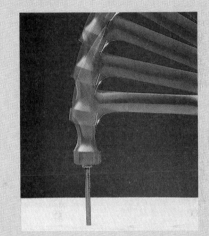

Features

Laboratory Investigations
Measuring Constant Speed   30
Will an Elephant Fall Faster Than a Mouse?   56
A Cartesian Diver   78
Up, Up, and Away!   102
Relating Mass, Velocity, and Kinetic Energy   126

Activity: Discovering
Marble Motion   19
Newton's Third Law of Motion   45
Science and Skydiving   50
A Plumber's Magic   69
An Archimedean Trick   72
Rolling Uphill   74
Follow the Bouncing Ball   123

Activity: Doing
Star Gazing   13
Demonstrating Inertia   43
Science and the Leaky Faucet   51
Faster Than a Speeding Snowball?   52
Air Pressure   64
To Float or Not to Float   73
Work and Power   87
Levers   97
Compound Machines   101
Energy in the News   109
Mixing It Up   118

Activity: Calculating
Move That Barge   44
It Takes Work to Catch a Flight   85
Computing Kinetic Energy   113

Activity: Thinking
Momentum   28
Simple Machines Around You   100
A Day at the Amusement Park   117

Activity: Writing
Race Around the World   23
A View of the World   47

Activity: Reading
Trotting Into Your Heart   15
Just Floating Away   20
Voyage to the Moon   53
Science in a Chocolate Factory   90

Problem Solving
In Search of Buried Treasure   21
All in a Day's Work   49
Attack of the Shower Curtain   75
How Energy Conscious Are You?   120

Connections
In a Flash   29
Which Way Is Up?   55
What a Curve!   77
The Power of Nature   88
Our Energetic World   124

Careers
Air-Traffic Controller   26
Industrial Designer   54
Solar Engineer   122

CONCEPT MAPPING

Throughout your study of science, you will learn a variety of terms, facts, figures, and concepts. Each new topic you encounter will provide its own collection of words and ideas—which, at times, you may think seem endless. But each of the ideas within a particular topic is related in some way to the others. No concept in science is isolated. Thus it will help you to understand the topic if you see the whole picture: that is, the interconnectedness of all the individual terms and ideas. This is a much more effective and satisfying way of learning than memorizing separate facts.

Actually, this should be a rather familiar process for you. Although you may not think about it in this way, you analyze many of the elements in your daily life by looking for relationships or connections. For example, when you look at a collection of flowers, you may divide them into groups: roses, carnations, and daisies. You may then associate colors with these flowers: red, pink, and white. The general topic is flowers. The subtopic is types of flowers. And the colors are specific terms that describe flowers. A topic makes more sense and is more easily understood if you understand how it is broken down into individual ideas and how these ideas are related to one another and to the entire topic.

It is often helpful to organize information visually so that you can see how it all fits together. One technique for describing related ideas is called a **concept map**. In a concept map, an idea is represented by a word or phrase enclosed in a box. There are several ideas in any concept map. A connection between two ideas is made with a line. A word or two that describes the connection is written on or near the line. The general topic is located at the top of the map. That topic is then broken down into subtopics, or more specific ideas, by branching lines. The most specific topics are located at the bottom of the map.

To construct a concept map, first identify the important ideas or key terms in the chapter or section. Do not try to include too much information. Use your judgment as to what is really important. Write the general topic at the top of your map. Let's use an example to help illustrate this process. Suppose you decide that the key terms in a section you are reading are School, Living Things, Language Arts, Subtraction, Grammar, Mathematics, Experiments, Papers, Science, Addition, Novels. The general topic is School. Write and enclose this word in a box at the top of your map.

SCHOOL

Now choose the subtopics—Language Arts, Science, Mathematics. Figure out how they are related to the topic. Add these words to your map. Continue this procedure until you have included all the important ideas and terms. Then use lines to make the appropriate connections between ideas and terms. Don't forget to write a word or two on or near the connecting line to describe the nature of the connection.

Do not be concerned if you have to redraw your map (perhaps several times!) before you show all the important connections clearly. If, for example, you write papers for Science as well as for Language Arts, you may want to place these two subjects next to each other so that the lines do not overlap.

One more thing you should know about concept mapping: Concepts can be correctly mapped in many different ways. In fact, it is unlikely that any two people will draw identical concept maps for a complex topic. Thus there is no one correct concept map for any topic! Even

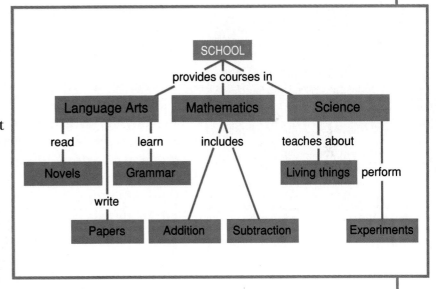

though your concept map may not match those of your classmates, it will be correct as long as it shows the most important concepts and the clear relationships among them. Your concept map will also be correct if it has meaning to you and if it helps you understand the material you are reading. A concept map should be so clear that if some of the terms are erased, the missing terms could easily be filled in by following the logic of the concept map.

MOTION, FORCES AND ENERGY

▲ Although this volcanic eruption may look like a fireworks display, it is actually an extremely violent release of energy.

Poised high upon the edge of a jagged rock, the cheetah seems almost motionless. Yet every muscle is tensed and ready to spring into action. A gentle breeze blows the familiar scent of an antelope toward the cheetah's quivering nostrils. Its head turns. The prey is in sight. The cheetah silently readies itself for attack, while the unsuspecting antelope calmly grazes on the plain just below.

With one swift leap, the cheetah pounces forcefully on the ground and sets into motion. The antelope takes off fast and furiously to spare itself an untimely death at the paws of the cheetah. The powerful cheetah can run as fast as 155 kilometers per hour. The antelope seems doomed. Speeding across the grassy plain with the chee-tah at its tail, the antelope suddenly darts into the bush. The cheetah lags behind. The antelope escapes— this time. A meal of antelope would have given the cheetah its fill of energy.

An anxious cheetah cub awaits an opportunity to pounce on and capture a handsome meal.

CHAPTER

1 What Is Motion?
2 The Nature of Forces
3 Forces in Fluids
4 Work, Power, and Simple Machines
5 Energy: Forms and Changes

But for tonight, the cheetah will go hungry.

The interaction of forces that made this motion-packed scene possible is not limited to the jungle. Almost everything you do every day of your life involves the same basic characteristics of motion, forces, and energy that spared this antelope's life. In this book you will learn all about motion and forces and the relationship between them. You will also learn how forces are altered by machines to make them more helpful to you. Finally, you will learn about energy and its role in motion, forces, and machines.

This pipeline wave may give the surfer the ride of his life or a crushing blow. Where do you think the wave gets its energy?

Discovery *Activity*

Tabletop Raceway

1. On a smooth surface such as a floor or long tabletop, make a ramp by placing one end of a sturdy piece of cardboard on one textbook. Make another ramp using three or four textbooks.

2. Release a small toy car from the top of the low ramp. Do not push it. Observe how far and how fast the car travels.

3. Repeat step 2 using the high ramp.

4. Tape a stack of washers to the car. Repeat steps 2 and 3. Observe the movement of the car.

   What effect does the height of the ramp have on the movement of the car?

   How would the movement of the car have changed if you had pushed it down the ramp?

   How do the washers affect the movement of the car?

   ■ Is it more difficult to stop a heavily loaded truck or a compact car?

What Is Motion?

Guide for Reading

After you read the following sections, you will be able to

1–1 Frames of Reference

■ Recognize that movement is observed according to a frame of reference.

1–2 Measuring Motion

■ Describe motion and calculate speed and velocity.

1–3 Changes in Velocity

■ Relate acceleration to motion and calculate acceleration.

1–4 Momentum

■ Describe momentum.

The eyes of the crowd are fixed on the sleek, dazzling skier as he sweeps down the ski-jump track at 100 kilometers per hour. Reaching the bottom of the track, he leaps into the air. The icy wind lashes at his face. Even with goggles on, he is blinded by the glare of the snow on the mountainside below—far below.

Then, in an attempt to defy gravity, he leans forward. His body and skis take the form of an airplane wing as he rides the wind farther. Finally, his skis make contact with the snow-covered landing area. The snow flies up in his face in two streams. The sound of the cheering crowd mingles with the sound of the wind. His ride is over.

The scene is the Winter Olympics. And the ski jumper has just flown 117 meters through the air to set a new record for distance. He owes his gold medal not only to courage and years of training but also to an understanding of motion.

In this sense, ski jumping is not merely a sport; it is also a science. Learning about the science of motion may not earn you an Olympic medal, but it can be a leap into adventure and discovery.

Journal *Activity*

You and Your World Recall an experience in which you were moving at a fairly rapid speed. Perhaps you were running, bicycling, or riding in a car, train, or airplane. Describe the situation in detail in your journal. Explain how you felt speeding up from rest and then slowing down again to stop. Tell how the speed affected your movement. How would your movement have been different if your speed was slower?

◀ *Although this ski jumper seems to be floating in midair, his body is really in motion. You would quickly observe this fact if you were watching the ski jump from the ground below.*

1–1 Frames of Reference

The conductor blows the whistle, and your train pulls away from the station. As the train picks up speed, the people on the platform watch you whiz by them. But to the person sitting next to you on the train, you are not moving at all. How can this be? Are you or are you not moving? Actually, you are doing both! The answer to the question depends on the background with which the observer is comparing you.

The people on the platform are comparing you with the Earth. Because the Earth is not moving out from under their feet, you appear to be moving. The person sitting next to you is comparing you with the train. Because you are moving with the train (the train is not moving out from under you), you do not appear to be moving when compared with the train. If, however, the person on the train had compared you with a nearby tree or the ground, you would have appeared to move.

Whenever you describe something that is moving, you are comparing it with something that is assumed to be stationary, or not moving. The background or object that is used for comparison is called a **frame of reference.** All movement, then, is described relative to a particular frame of reference. For the people on the platform, the frame of reference is the Earth. For the person sitting next to you, the frame

Figure 1–1 *From your frame of reference, the sun seems to be dropping below the horizon in southern Spain. But is the sun really falling or are you moving?*

Figure 1–2 *These Navy fighter jets are traveling at tremendous speeds relative to an observer on the ground. But because they are moving at the same speed, they are not moving relative to one another. This enables them to hook up and refuel in midair.*

of reference is the train. Perhaps the term frame of reference sounds new to you. But it is an idea you use often in your daily life as you describe different movements. The frame of reference you use depends on the type of movement and the position from which you are observing.

The fact that movement is related to a frame of reference is often used in movies to achieve certain effects. Sometimes an actor stays in one place and just the background moves. On the screen it looks as if the actor is moving. This is because your frame of reference is the background. You have assumed that the background is the stationary object—which it is most of the time.

No frame of reference is more correct than any other. If you are riding on a train, you may describe movement as if the train were your frame of reference. However, the train is moving relative to the Earth. So, if you use the trees and the ground as your background, the Earth could become your frame of reference. But the Earth is moving around the sun. Thus the sun could be your frame of reference. Even the sun is moving as part of the galaxy. So you see, everything in the universe is moving. There is no frame of reference that is truly not moving relative to all other frames of reference. What is not moving in one frame of reference is moving in another. But all movement is described according to some frame of reference. The most common frame of reference is the Earth, but no single frame of reference is "correct" in any situation.

Figure 1–3 *No need to turn this photograph right-side-up. Fred Astaire is dancing on the ceiling, or so it seems. How is this effect achieved in the movies?*

ACTIVITY

DOING

Star Gazing

The Big Dipper is a dipper-shaped group of stars familiar to most people. If you have ever tried to locate the Big Dipper in the nighttime sky, you know that it is not always in the same place. In January, you may look out your window in one direction to find it. In June, however, you will have to look in another direction.

Obtain star charts for 12 months in a row. Observe how the stars are in different locations in the course of the year. Is the motion a result of the movement of the Earth or of the stars? What does your description of motion depend on?

1–1 Section Review

1. What is a frame of reference?
2. What is the most common frame of reference?
3. How is it possible that an actress can be shown dancing on the ceiling in a movie?

Critical Thinking—*Applying Concepts*

4. Suppose you are standing on a sidewalk and your friend rides past you on her skateboard. Which one of you is moving relative to the Earth? Are you moving relative to your friend?

1–2 Measuring Motion

A cool autumn breeze sends a leaf on a spiraling journey to the ground. An army of ants marches past your feet on their way to your fallen ice cream. A meteor leaves a brilliant streak of light in its path as it hurtles through the atmosphere. In each one of these examples, something is changing position, or moving from one place to another. And although it might take the ants a while to reach their destination whereas the meteor is gone in a few blinks of the eye, all of these movements take place over some particular amount of time. **A change in position in a certain amount of time is motion.** When you say that something has moved, you are describing **motion.** But always remember that when you describe movement, or motion, you are comparing it with some frame of reference.

Speed

Suppose that at one instant runners are poised at the starting blocks ready for a race. Seconds later, the winner breaks the tape at the finish line. The runners got from the starting blocks to the finish line because they moved, or changed their position.

Figure 1– 4 *Just as the runners move from one place to another during their race, so do the small ants as they carry bits of a leaf back to their home. Which photo shows movement at a greater speed?*

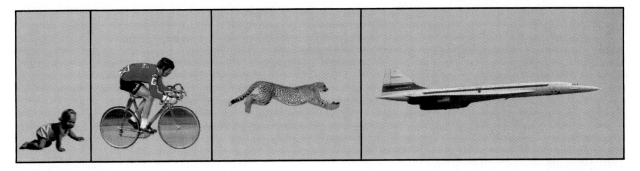

| 1 km/hr | 80 km/hr | 110 km/hr | 3600 km/hr |
| Baby crawling | Cyclist | Cheetah | Concorde SST |

And they did so in a certain amount of time. So you can be sure that the runners were in motion. But to better describe the motion of the runners, you need to know the distance they traveled and how long it took them to travel that distance (that is, to reach the finish line). Distance is the length between two places. In the metric system, distance is measured in units called meters (m) and kilometers (km). One kilometer is equal to 1000 meters. How many meters are there in 10 kilometers?

If you know the distance the runners traveled and the time it took them to travel that distance, then you can determine how fast each runner moved. In other words, you can calculate the **speed** of each runner. You probably use the words fast and slow often to describe motion. But what you may not realize is that by using these words, you are actually describing speed. **Speed is the rate at which an object moves.** The faster a runner's rate of motion, the faster the runner's speed.

You can find the speed of an object by dividing the distance it traveled by the time it took to travel that distance:

$$\text{Speed} = \frac{\text{Distance}}{\text{Time}}$$

Using this formula, you can see that if two runners ran the same distance, the runner who took the longer time must have run at a slower speed. The runner who took the shorter time must have run at a faster speed. And as you know, only the runner who ran faster won the race. If, on the other hand, two runners took the same amount of time but one runner ran a longer distance, that runner must have run at a faster speed.

Figure 1–5 *You can compare the speeds of some common objects on this scale. Where do you think your walking speed would fit?*

ACTIVITY
READING

Trotting Into Your Heart

When you think of speed, the image of a beautiful sleek horse may come to mind. That same image must have come to Walter Farley's mind because he wrote an entire collection of books about a black stallion. Read *The Black Stallion* by Walter Farley and enjoy the adventures of a very special horse. And if you get hooked on his stories, which you may easily do, go ahead and indulge in the rest of the series.

Distance is usually measured in meters or kilometers. Time is usually measured in seconds or hours. So the unit of speed is often meters per second (m/sec) or kilometers per hour (km/hr). This idea should be pretty familiar to you. After all, if someone asked you for the speed of a car, would you say 80 kilometers? No! You would

Figure 1–6 *Microscopic red blood cells crowd together as they move through the body at a particular speed. This huge glacier moves through part of Alaska at a different speed—obviously slower!*

say 80 kilometers per hour. This means that traveling at this speed the car will go a distance of 80 kilometers in a time of one hour.

Speed is not used to describe only runners, cars, or trains. Anything that is changing its position has speed. This includes ocean currents, wind, glaciers, the moon, and even the Earth. For example, from the sun's frame of reference, the Earth is orbiting the sun at an average speed of about 30 kilometers per second, or 30 km/sec.

CONSTANT SPEED So far, we have been discussing the motion of objects whose speed is the same throughout their movement. Their speed does not change. Speed that does not change is called constant speed. When you calculate the speed of an object traveling at constant speed, you are figuring out its speed at every point in its path. Let's see how this works. Look at Figure 1–7. This is a

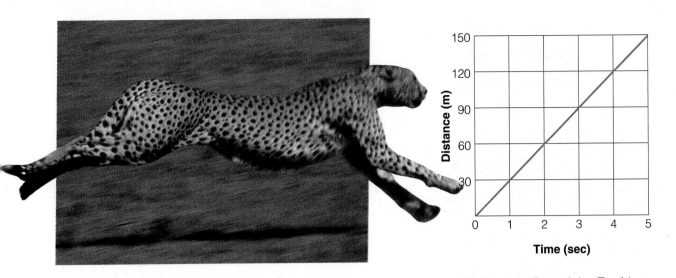

Figure 1–7 *One of the Earth's swiftest land animals, the cheetah can attain great speeds for short periods of time. The graph shows the distance the cheetah travels as a function of time. How do you know the cheetah ran at a constant speed?*

distance–time graph showing several seconds of a cheetah's motion. Distance is plotted on the vertical, or Y, axis. Time is plotted on the horizontal, or X, axis. According to the graph, how many meters did the cheetah travel after 1 second? You are right if you said 30 meters. The cheetah's speed was 30 m/1 sec, or 30 m/sec. After 3 seconds, the cheetah traveled 90 meters. So its speed was 90 m/3 sec, or 30 m/sec. The cheetah's speed did not change. When you divide the total distance by the total time, 150 m/5 sec, you get 30 m/sec. Thus, for constant speed, total distance divided by total time gives the speed for every point in the cheetah's path because

Sample Problem	At what speed did a plane fly if it traveled 1760 meters in 8 seconds?
Solution	
Step 1 Write the formula.	**Speed = Distance/Time**
Step 2 Substitute given numbers and units.	**Speed = 1760 meters/8 seconds**
Step 3 Solve for the unknown.	**Speed = 220 meters/second (m/sec)**
Practice Problems	**1.** A car travels 240 kilometers in 3 hours. What is the speed of the car during that time?
	2. The speed of a cruise ship is 50 km/hr. How far will the ship travel in 14 hours?

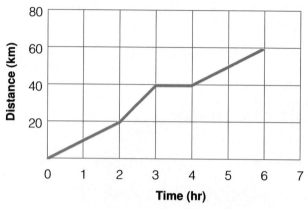

Figure 1–8 *Riding merrily along on a lovely spring day, the cyclist travels quite a distance. The graph shows how her speed changed during the course of the trip. During which hours was her speed the greatest?*

Activity Bank

Flying High, p.138

Figure 1–9 *If you have ever walked into the wind, you know you expend more energy than you do if you walk with the wind at your back. Why?*

the speed of the cheetah is the same at every point. Notice that the distance–time graph for constant speed is a straight line.

AVERAGE SPEED The speed of a moving object is not always constant. In fact, this is usually the case. Suppose a cyclist takes a long ride. She begins slowly for the first two hours and picks up speed for the third hour. After three hours, she takes a break before finishing the final two hours of her ride. Clearly, her speed changes many times throughout her journey. Dividing the total distance by the total time does not tell you her speed for every point of her journey. Instead it gives you her average speed. What is the average speed of the cyclist in Figure 1–8? Actually, the formula you just learned for calculating speed—distance/time—always gives you the average speed. For things that move at constant speed (such as the cheetah in the previous example), the speed at any point is the same as the average speed.

Velocity

In addition to speed, another characteristic is needed to describe motion. It is the direction in which an object moved. Did it go east? West? North? South? To describe both the speed and the direction of motion, **velocity** is used. **Velocity is speed in a given direction.** Suppose a runner moves eastward at 10 m/sec. Her *speed* is 10 m/sec. Her *velocity*, however, is 10 m/sec east. If the runner was moving westward, her speed would be the same, but her velocity would change.

You may not think that there is much difference between speed and velocity. But the direction given by velocity is very important. Airplane pilots and air-traffic controllers use velocity measurements to fly and land airplanes. It would be dangerous to know only that one airplane is taking off at a speed of 100 km/hr and that another airplane is landing at 150 km/hr. Instead, air-traffic controllers must know the airplanes' directions so that the airplanes will not head directly into each other. Weather forecasters must know the velocity of air masses to predict the weather correctly. And anyone traveling from one city to another knows that it is not only the speed that counts, but also the direction!

COMBINING VELOCITIES Suppose you are rowing a boat downstream at 16 km/hr. Would it surprise you to learn that you are actually going faster than 16 km/hr? How is this possible? The river is also moving, say at 10 km/hr. Since you are rowing

ACTIVITY
DISCOVERING

Figure 1–10 *These adventurous white-water rafters are spared a little work by the river. Why does the river make them move much faster than they are rowing?*

Figure 1–11 *A rocket launched in the same direction as the Earth rotates gets an added boost. Without this boost, Space Shuttles would have to attain much greater speeds in order to leave the Earth's atmosphere and enter space. How much of a boost does a rocket get?*

1800 km/hr

39,200 km/hr

41,000 km/hr

downstream, you are going in the same direction as the river. The two velocities combine. So you are moving at 26 km/hr. Velocities that have the same direction are added together. You must use subtraction when combining velocities in opposite directions. For example, if you were rowing 16 km/hr upstream, you would be going 16 km/hr – 10 km/hr, or 6 km/hr.

This idea of combining velocities is very important, especially for launching rockets and flying planes. Space engineers launch rockets in the same direction as the Earth rotates. The speed of the Earth's rotation is about 1800 km/hr. Thus the rocket gets an added boost of 1800 km/hr to its speed.

1–2 Section Review

1. Define motion. Give an example.
2. How is the speed of an object calculated?
3. What is constant speed? How does it compare with average speed?
4. What quantity gives both the speed and the direction of an object?

Connection—*You and Your World*
5. It is 2:00 PM. The local weather bureau is tracking a violent storm that is traveling eastward at 25 km/hr. It is 75 kilometers west of Suntown, USA. If everyone works until 5:00 PM, will the residents of Suntown be able to get home safely before the storm hits?

PROBLEM Solving

In Search of Buried Treasure

The *Sea Queen,* a majestic research ship, drops anchor in the area where a great treasure is believed to have been lost on the ocean floor. Here, however, the ocean floor is far too deep for divers to explore. So instead, the scientists on board send out a series of sound waves at different locations around the ship. Sound waves travel at 1.5 km/sec in ocean water. When a sound wave reaches the ocean bottom, it bounces back to the ship, where it is recorded. Each sound wave travels the same distance on the way up from the ocean floor as it does on the way down.

After a long day, the researchers have gathered

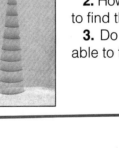

many pages of data describing the sound waves. Now it is time to call in an expert to interpret the information. And guess what—you are the expert! It will be your job to figure out the depth of the ocean at each point in the area surrounding the ship.

Organizing Resources and Information

1. What other measurement do you need to do this?

On board is a map of the ocean floor in that area. The map was made on the basis of measurements taken before the treasure was lost.

2. How can you use the map to find the treasure?

3. Do you think you will be able to find the lost treasure?

1-3 Changes in Velocity

Have you ever ridden on a roller coaster? You are pulled up to the top of the first hill at constant speed. But as you roll down the other side, your speed rapidly increases. At the bottom of the hill you make a sharp right turn, changing your direction once again. Then your speed rapidly decreases as you climb the second hill. On a roller-coaster ride, you experience rapid changes in velocity. Remember, velocity measures both speed and direction.

The rate of change in velocity is known as acceleration. If something is accelerating, it is

Guide for Reading

Focus on these questions as you read.

▶ *What is the relationship between velocity and acceleration?*

▶ *How is acceleration calculated?*

Figure 1–12 *Up, down, and around go the roller-coaster cars. If a car begins its final descent at 4 km/hr and zooms down the hill in 5 seconds, what other information do you need to calculate its final acceleration?*

speeding up, slowing down, or changing direction. The **acceleration** of an object is equal to its change in velocity divided by the time during which this change occurs. The change in velocity is the difference between the final velocity and the original velocity:

$$\text{Acceleration} = \frac{\text{Final Velocity} - \text{Original Velocity}}{\text{Time}}$$

Acceleration, like speed, tells you how fast something is happening—in this case, how fast velocity is changing. And like velocity, acceleration has direction. A car on an entrance ramp to a highway begins at rest and gradually increases its speed in the

---

Sample Problem
A roller coaster's velocity at the top of a hill is 10 meters/second. Two seconds later it reaches the bottom of the hill with a velocity of 26 meters/second. What is the acceleration of the roller coaster?

Solution

Step 1   Write the formula. $$\text{Acceleration} = \frac{\text{Final Velocity} - \text{Original Velocity}}{\text{Time}}$$

Step 2   Substitute given numbers and units. $$\text{Acceleration} = \frac{26 \text{ meters/second} - 10 \text{ meters/second}}{2 \text{ seconds}}$$

Step 3   Solve for the unknown. $$\text{Acceleration} = \frac{16 \text{ meters/second}}{2 \text{ seconds}}$$

$$\text{Acceleration} = 8 \text{ meters/second/second (m/sec/sec)}$$

Practice Problems

1. A roller coaster is moving at 25 m/sec at the bottom of a hill. Three seconds later it reaches the top of the next hill, moving at 10 m/sec. What is the deceleration of the roller coaster?

2. A car is traveling at 60 km/hr. It accelerates to 85 km/hr in 5 seconds. What is the acceleration of the car?

direction of the highway. The car is accelerating in the direction of the highway.

You can determine the unit of acceleration by looking at the formula. The change in velocity is measured in kilometers per hour or meters per second. Time is measured in hours or seconds. So acceleration is measured in kilometers per hour per hour (km/hr/hr) or meters per second per second (m/sec/sec). This means that if an object is accelerating at 5 m/sec/sec, each second its velocity increases by 5 m/sec. The speed of the object is 5 m/sec greater each second.

If there is a decrease in velocity, the value of acceleration is negative. Negative acceleration is called deceleration. When a roller coaster climbs a hill, it decelerates because it is slowing down. Can you think of another example of deceleration?

The data table in Figure 1–13 is a record of a professional drag-strip race. The driver had traveled a distance of 5 meters after the first second. The

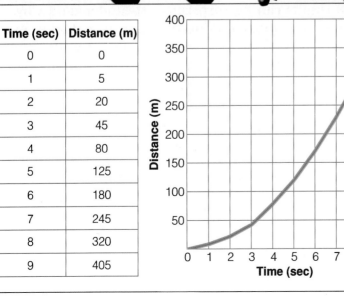

Time (sec)	Distance (m)
0	0
1	5
2	20
3	45
4	80
5	125
6	180
7	245
8	320
9	405

Figure 1–13 *The data from a professional drag race are shown on the left. A distance–time graph of the racer's motion is shown on the right. What is the acceleration of the race car?*

Figure 1–14 *Although these photos may make you dizzy, they each illustrate important properties of circular motion. Why are both the gymnast and a passenger on the Ferris wheel experiencing acceleration?*

distance covered in the next second was 15 meters (20 m – 5 m). By the end of four seconds, the driver had traveled 80 meters. Figure 1–13 on page 23 also shows a distance–time graph of the racing car's motion. The graph is a curve rather than a straight line. A distance–time graph for acceleration is always a curve. How far did the driver travel in the first five seconds of the race? In the last five seconds?

Circular Motion

Acceleration and deceleration are easy to recognize when the motion is in a straight line. After all, motion in a straight line does not involve a change in direction. It involves only a change in speed. And it is rather easy to recognize an object speeding up or slowing down. When the path of motion is curved, however, the results can be surprising. To understand why, you must remember that acceleration is a change in velocity. And velocity expresses direction as well as speed. In circular motion, the velocity is continuously changing because direction is continuously changing. An object in circular motion is accelerating even though its speed may be constant.

You experience circular motion in many common activities. When you ride a Ferris wheel, pedal a bike on a curved track, or travel in a car turning a corner, you are constantly changing direction. So you are accelerating. In fact, because the Earth is continuously rotating, you are constantly accelerating—even when you are fast asleep!

1–3 Section Review

1. What is acceleration?
2. What is another name for negative acceleration?
3. Why does an object traveling in a circular path at constant speed accelerate?
4. A car at a stoplight has a velocity of 0 km/hr. Three seconds after the light turns green, the car has a velocity of 30 km/hr. What is the acceleration of the car?

Critical Thinking—*Making Calculations*

5. Despite his mother's warnings, Timothy was playing ball in the house when his ball bounced out the window. A freely falling body accelerates at about 10 m/sec/sec. What is the velocity of Timothy's ball after 2 seconds? *Hint:* What is its original downward velocity?

Figure 1–15 *Powering the spinning wheel with her foot, this woman takes advantage of circular motion. Can you think of other devices that use circular motion?*

1–4 Momentum

The 100-kg fullback runs up the middle of the football field. Suddenly he collides with a 75-kg defensive back running toward him. The more massive fullback is thrown back two meters! How can a 75-kg defensive back stop a 100-kg fullback?

The answer is that the defensive back has more **momentum.** All moving objects have momentum. And the more momentum an object has, the harder it is to stop. **Momentum depends on the mass of the object and the velocity with which it is traveling.** If either of these measurements is large, the object will have a large momentum.

Momentum is equal to the mass of an object multiplied by its velocity:

$$\textbf{Momentum} = \textbf{Mass} \times \textbf{Velocity}$$

Although he has less mass, the defensive back has more momentum because he is moving faster than the fullback. His greater velocity makes up for his smaller mass. If both players had the same velocity,

Guide for Reading

Focus on these questions as you read.

▶ *How is momentum related to velocity and mass?*

▶ *What does it mean that momentum is conserved?*

Figure 1–16 *It has been said that football is a game of momentum. What must be true of the smaller player's velocity if his momentum is enough to stop the larger player?*

who would have more momentum? You can determine the unit of momentum by looking at the formula. Mass is often measured in kilograms and velocity in meters per second. So the unit of momentum is usually kilogram-meters per second (kg-m/sec). And momentum is in the same direction as the velocity of the object.

A train has a large momentum because of its mass. That is probably obvious to you. But what do you think about the momentum of a bullet fired from a rifle? Do you think it is small? Do not be fooled by the small mass of a bullet. Bullets are fired at extremely high velocities. A bullet has a large momentum because of its velocity. What is the momentum of a bullet before it is fired from a rifle?

Why do you think it is harder to stop a car moving at 100 km/hr than it is to stop the same car moving at 25 km/hr? You are right if you said the car moving at 100 km/hr has more momentum. A car having greater momentum requires a longer distance in which to stop. The stopping distance of a car is directly related to its momentum. People who design roadways must take this into account when determining safe stopping distances and speed limits.

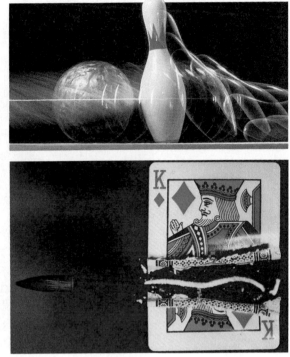

Figure 1–17 *Although the King of Diamonds may not appreciate the importance of momentum, the bullet is able to cut right through the card because of this physical phenomenon. Similarly, the momentum of the bowling ball enables it to knock the pins down. Why must the engines of supertankers be shut off several kilometers before they need to stop?*

Conservation of Momentum

If you have ever played billiards, you may know about an important property of momentum. If you send a moving ball into a stationary ball, you can cause the stationary ball to move and the moving ball to become stationary. This is because the momentum of the moving object is transferred to the stationary object when the two objects collide. None of the momentum is lost. **The total momentum of any group of objects remains the same unless outside forces act on the objects.** This is what we mean when we say that momentum is conserved. One object may lose momentum, but the momentum lost by this object is gained by another. Momentum is always conserved.

Now suppose that you send two moving billiard balls into each other. After they hit, both balls are still moving. This means that neither ball transferred all of its momentum to the other. But as you just learned, momentum is conserved. So the total momentum of the billiard balls before they hit and after they hit must be the same. Although their

Figure 1–18 *Like a long jumper who uses her momentum to carry her a great distance, an agile frog also takes advantage of momentum. Why do long jumpers run before jumping?*

individual momentums may change, because of either a change in speed or mass, the total momentum does not change. For example, if one ball speeds up after they hit, the other must slow down.

There are many common examples of conservation of momentum. The momentum of a baseball bat is transferred to the ball when the bat and the ball meet. The more momentum the bat has, the more momentum is transferred to the ball. The act of throwing an object off a boat causes the boat to move in the opposite direction. The more massive the object and the faster it is thrown, the faster the boat will move away. Why do you think a pitcher winds up before throwing the baseball? In each of these situations, the total momentum is conserved.

ACTIVITY

THINKING

Momentum

Place the following objects in the correct order from the lowest to the highest momentum. Assume that all of the objects are moving at their maximum velocity.

freight train   mosquito
bullet   automobile
Space Shuttle

1–4 Section Review

1. What is momentum?
2. How is momentum conserved?
3. What is the momentum of an 0.30 kg bluejay flying at 17 m/sec?
4. Which object has more momentum: a car traveling at 10 km/hr or a baseball pitched at 150 km/hr? Explain your answer.

Critical Thinking—*Making Inferences*
5. When a person jumps from a tree to the ground, what happens to the momentum of the person upon landing on the ground?

In a Flash

Think about some of the photographs you or your family have taken. Most likely, they are of events or people who are not moving—a snowcapped mountain, a famous statue, some friends posing. But have you ever tried to photograph things that are in motion—perhaps friends who moved just as you pushed the button on your camera? If so, you know that the photograph comes out blurry. So you might wonder how cameras can photograph moving objects—particularly those that move too quickly even for the human eye to see.

Cameras depend on light. When light bounces off an object it will form a picture on a material that is sensitive to light. That material is film. This picture can then be chemically processed into a photograph. A device on a camera called a shutter controls the length of time that the film is exposed to light. If the subject moves at any time while the film is exposed to light, the movement will be recorded as a blur.

By decreasing the length of time the shutter is open to only fractions of a second, photographers are able to take sharp pictures of moving subjects. Remember that motion involves a change in position during a certain amount of time. When the speed with which the shutter opens and closes is increased, the time segment studied becomes so small that the change in position is too small to be recorded as a blur. The camera actually catches a tiny segment of motion.

At a setting of 1/1000 of a second, the shutter is open for such a short time that the motion of a race car appears to be "stopped." To freeze the beating of an insect's wings needs an even shorter exposure. Even at 1/1000 of a second, the wings are a blur. To record this type of motion, high-speed cameras capable of exposures 10 or 20 times shorter have been developed.

Photography has become a universal means of communication and a valuable tool in many fields. Photographs made at high speeds are very important in science and technology. This is because they record a phase of a fast event or photograph a rapid sequence of events that can be slowed down for study. The ability to photograph moving subjects has opened up a whole new world of study and has shed light on important aspects of the physical world. Some wonderful contributions to science have been made by creative and ambitious shutterbugs.

Laboratory Investigation

Measuring Constant Speed

Problem

What is the shape of a distance–time graph of constant speed?

Materials (per student)

pencil
graph paper
metric ruler

Procedure

1. The illustration on this page represents a series of flash shots taken of a dry-ice puck sliding across the floor. The time be-tween each flash is 0.1 second. Study the illustration carefully.

2. Copy the sample data table on a piece of graph paper.

3. Position the 0-cm mark of the metric ruler on the front edge of the first puck. This position will represent distance 0.0 cm at time 0.0 second. Record these data in your data table.

4. Without moving the ruler, determine the distance of each puck from the first one.

5. Record each distance to the nearest 0.1 cm in your data table.

Observations

Time (sec)	Distance (cm)
0.0	0.0
0.1	
0.2	
0.3	
0.4	
0.5	
0.6	

Make a distance–time graph using the data in your table. Plot the distance on the vertical, or Y, axis and the time on the hor-izontal, or X, axis.

Analysis and Conclusions

1. What is the shape of the graph?
2. Is the speed constant? Explain your an-swer.
3. Calculate the average speed.
4. How will the graph change as time goes on?
5. **On Your Own** Suppose you are ice skat-ing around a rink at a constant speed. Then you get tired so you stop moving your feet and glide along the ice. How would your distance–time graph look?

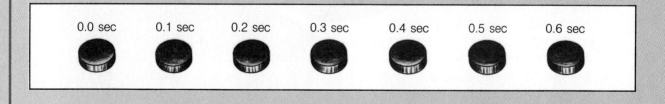

Summarizing Key Concepts

1–1 Frames of Reference

▲ All movement is compared with a background that is assumed to be stationary. This background is called a frame of reference.

▲ An object that is stationary in one frame of reference may be moving in another frame of reference. Any frame of reference can be chosen to describe a given movement, but the most common frame of reference is the Earth.

1–2 Measuring Motion

▲ Motion involves a change in position during a certain amount of time. The characteristics of position and time are used to measure motion.

▲ The rate at which an object moves is speed. Any object that is changing its position has speed. Speed can be determined by dividing the distance traveled by the time taken to travel that distance.

▲ Speed that does not change is called constant speed. For an object moving at constant speed, the speed at any point is the same as the average speed. For an object whose speed varies, you calculate the average speed.

▲ Speed in a given direction is velocity.

▲ Velocities that have the same direction combine by addition. Velocities that have opposite directions combine by subtraction.

1–3 Changes in Velocity

▲ Acceleration is the rate of change in velocity. It is equal to the change in velocity divided by the time it takes to make the change.

▲ An object that is accelerating is speeding up, slowing down, or changing direction.

▲ Negative acceleration is also known as deceleration.

▲ Circular motion always involves acceleration because the object's direction is constantly changing.

1–4 Momentum

▲ Momentum is equal to the mass of an object multiplied by its velocity. An object with a large momentum is very difficult to stop.

▲ The total momentum of any group of objects remains the same unless outside forces act on the objects.

Reviewing Key Terms

Define each term in a complete sentence.

1–1 Frames of Reference
frame of reference

1–2 Measuring Motion
motion
speed
velocity

1–3 Changes in Velocity
acceleration

1–4 Momentum
momentum

Chapter Review

Content Review

Multiple Choice

Choose the letter of the answer that best completes each statement.

1. All movement is compared with a
   a. car.
   b. frame of reference.
   c. tree.
   d. train.

2. The most commonly used frame of reference is the
   a. sun.
   b. Earth.
   c. moon.
   d. ocean.

3. A change in position relative to a frame of reference is
   a. motion.
   b. momentum.
   c. acceleration.
   d. direction.

4. The rate at which an object changes position is called
   a. distance.
   b. acceleration.
   c. speed.
   d. momentum.

5. Velocity is speed and
   a. motion.
   b. mass.
   c. distance.
   d. direction.

6. If a motorboat travels 25 km/hr down a river whose velocity is 4 km/hr, what is the boat's actual velocity?
   a. 21 km/hr
   b. 29 km/hr
   c. 100 km/hr
   d. 6.2 km/hr

7. The rate of change of velocity is called
   a. speed.
   b. motion.
   c. momentum.
   d. acceleration.

8. A distance-time graph is a straight line for
   a. constant speed.
   b. acceleration.
   c. momentum.
   d. average speed.

9. An object traveling in circular motion is constantly changing
   a. speed.
   b. mass.
   c. distance.
   d. direction.

10. Momentum is mass times
    a. acceleration.
    b. velocity.
    c. motion.
    d. distance.

True or False

If the statement is true, write "true." If it is false, change the underlined word or words to make the statement true.

1. Motion must be measured relative to a <u>frame of reference</u>.
2. A change in position of an object is called <u>momentum</u>.
3. The measurement of how fast or slow something is traveling is <u>speed</u>.
4. An object whose speed does not change is traveling at <u>constant</u> speed.
5. The quantity that gives speed and direction is <u>momentum</u>.
6. Velocities in opposite directions combine by <u>subtraction</u>.
7. Acceleration is a change in speed or <u>direction</u>.
8. The measurement of how hard it is to stop an object (mass times velocity) is <u>acceleration</u>.

Concept Mapping

Complete the following concept map for Section 1–2. Refer to pages S6–S7 to construct a concept map for the entire chapter.

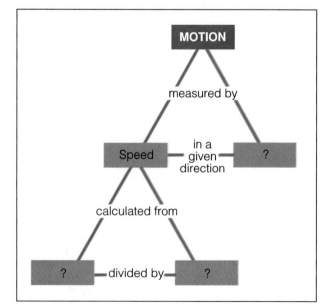

Concept Mastery

Discuss each of the following in a brief paragraph.

1. A car is traveling along the road at a moderate speed. One person describes the velocity of the car as 30 km/hr forward. Another person describes the car as moving 20 km/hr in reverse. Explain how both observers can be correct.

2. You are flying in an airplane whose speed is programmed to be 400 km/hr. However, the airplane is really traveling at 460 km/hr. Explain how this can be true.

3. Explain why you are being accelerated on a Ferris wheel moving at constant speed.

Critical Thinking and Problem Solving

Use the skills you have developed in this chapter to answer each of the following.

1. **Making calculations** Complete the following problems:
   a. What is the average speed of a jet plane that flies 7200 km in 9 hours?
   b. The speed of a cruise ship is 50 km/hr. How far will the ship travel in 14 hours?
   c. A car accelerates from 0 km/hr to 60 km/hr in 5.0 seconds. What is the car's acceleration? Watch your units!

2. **Applying definitions** Samantha ran 90 meters in 35 seconds to catch up with her dog. When she got to him, she played with him for 70 seconds before they walked back in 75 seconds. Did Samantha travel at constant speed? What was Samantha's average speed?

3. **Applying concepts**

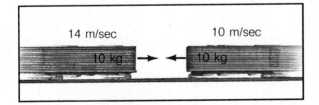

14 m/sec · 10 kg → ← 10 kg · 10 m/sec

   a. What is the momentum of the train car moving at 14 m/sec? Of the car moving at 10 m/sec? What is the total momentum of the system?
   b. If the two cars collide and stick together, what will be the direction of their resulting motion?

4. **Relating cause and effect** Use the following information to explain the launch of a rocket: Hot gases that escape from a rocket have a very small mass but a high velocity. As fuel is used up, the mass of the rocket decreases.

5. **Applying concepts** An old legend tells the story of a stingy man who never let go of his bag of coins. One winter's day, he slipped in the snow and suddenly found himself in the middle of a frozen pond. The ice on the pond was so smooth and slippery that he could not grab on to the ice to stand up. In fact, he could not get enough traction to move on it at all. What could he do to save himself?

6. **Using the writing process** The distance–time graph describes your walk to the local store and back home. Write a brief story describing your walk that would correspond to the graph.

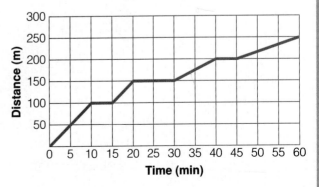

The Nature of Forces

Guide for Reading

After you read the following sections, you will be able to

2–1 What Is Force?

■ Describe the nature of force.

■ Compare balanced and unbalanced forces.

2–2 Friction: A Force Opposing Motion

■ Identify different types of friction.

2–3 Newton's Laws of Motion

■ Discuss Newton's three laws of motion.

■ Explain why Newton's laws of motion are important for describing common examples of motion.

2–4 Gravity

■ Describe the relationship between gravitational force, mass, and distance.

■ Compare weight and mass.

The year was 1665. Throughout London, schools and businesses had closed. The deadly bubonic plague raged through the city, causing twenty-two-year-old Isaac Newton to return to his mother's farmhouse in Woolsthorpe.

One day, Newton observed an apple falling from a tree. He began to wonder: Why does the apple fall down to the Earth?

During the next year, Isaac Newton proved that the force that pulls an apple to the ground is the same force that helps keep the moon in orbit around the Earth. He was also able to show that this force keeps the planets in their orbits around the sun. While Newton was making this profound discovery, he was also uncovering the secrets of light and color, and inventing a branch of mathematics called calculus. Incredibly, Newton accomplished all this in just 18 months!

Isaac Newton is considered the founder of modern physics and "one of the greatest names in the history of human thought." In this chapter, you will gain an appreciation for Newton and his contribution to science as you read about his beautifully simple explanation of forces and motion.

Journal *Activity*

You and Your World   Have you ever tried to pull something that just wouldn't budge? Maybe it was a stubborn dog avoiding a bath or a heavy piece of furniture. In your journal, describe a situation in which you pulled, or tried to pull, something. Include any details that made the job more or less difficult for you. What might have made your task easier?

◀ *Isaac Newton discovered the force that keeps the moons orbiting around Saturn and also holds you on the Earth.*

2–1 What Is Force?

Do you play baseball or tennis? Have you raked a pile of leaves or shoveled snow? Have you ever hammered a nail into a piece of wood or moved a large piece of furniture? How about something as simple as riding a bicycle, lifting this textbook, or opening a door? If so, you know that there is some type of motion involved in all of these activities. But what causes a tennis ball to suddenly zoom across the court or a bicycle to skid to a halt? The answer is **force.** In each of these activities, a force is involved. You are exerting a force on an object. And although you may not know it, the object is exerting a force on you! What is force? How is it related to motion?

A force is a push or pull. The wind pushes against the flag on a flagpole. A magnet pulls iron toward it. A jet engine pushes an airplane forward. The moon pulls on the oceans, causing the daily tides. A nuclear explosion pushes nearby objects outward with tremendous force. A negatively charged particle and a positively charged particle are attracted to each other. In each of these examples, a force is involved. **A force gives energy to an object, sometimes causing it to start moving, stop moving, or change direction.** For example, if you want to open a door, you exert a force on it to cause it to move. Increasing your force will make it move faster. If you want to stop the door from opening, you also exert a force. This time the force stops the motion of the door. And if you want to change the direction in which the door is moving, you must exert a force on it.

Figure 2–1 *Quite a force is required to send a soccer ball hurtling down a field. What is the source of the force?*

Figure 2–2 *Powerful ocean waters smash into coastal rocks all day. If you have ever been hit by ocean waves, you know just how forceful they can be.*

Figure 2–3 *The force exerted by a powerful pooch overcomes the opposing force of a reluctant child. Pulling a stubborn cow, however, may not be as enjoyable an adventure.*

Combining Forces

Have you noticed by now that most measurements involving motion—velocity, acceleration, and momentum, for example—include direction? Forces also act in a particular direction. Suppose you were trying to pull a wagon filled with rocks. To get the wagon to move, you would have to exert a force on the wagon and rocks. If your force was not large enough, you might ask a friend to help you. Your friend might pull with you or push from the back of the wagon. In either case, the two forces (yours and your friend's) would be exerted *in the same direction.* When two forces are acting in the same direction, they add together. The total force on the wagon would be the sum of the individual forces. When the total force on an object is in one direction, the force is called unbalanced. An unbalanced force changes the motion of an object.

If your friend pulled *in the opposite direction,* the forces would combine in a different way. When two forces act in opposite directions, they combine by subtraction. If one force was greater than the other force, the wagon would move in the direction of the greater force. And the total force on the wagon would be the difference between the individual forces. In this case, your friend would certainly not be helpful! What do you think would happen if your force and your friend's force were equal? When you subtracted one force from the other, you would be

Figure 2–4 *Moving this pumpkin is a hard job to do alone! So these two children are combining forces in the same direction. Are the children exerting a balanced or an unbalanced force?*

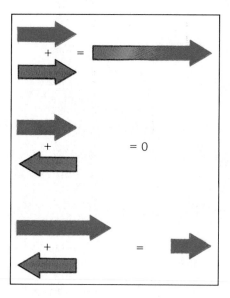

Figure 2–5 *Two forces can combine so that they add together (top), cancel each other (center), or subtract from each other (bottom).*

left with zero. This means that there would be no force acting on the object. The wagon would not move! Forces that are in opposite directions and equal in size are called balanced forces. When forces are balanced, there is no change in motion.

You should remember that in describing forces, a number value, a unit of measurement, and a direction must be given. It is helpful to think of forces as arrows. The length of the arrow shows the strength of the force. The head of the arrow points in the direction of the force. Using such arrows, you can tell what the resulting size and direction of combined forces will be.

2–1 Section Review

1. What is force?
2. How are forces related to motion?
3. What are unbalanced forces? Balanced forces?

Connection—*Life Science*
4. How is your heart able to produce a force? Why is this force vital to life?

Guide for Reading

Focus on these questions as you read.

▶ *What are the effects of friction on motion?*
▶ *What are three types of friction?*

2–2 Friction: A Force Opposing Motion

Have you ever tried to slide a piece of furniture, such as a desk, across a floor? If you have, you know that as you push, the rubbing of the desk against the floor makes it difficult to push the desk. This is because whenever two surfaces are touching, such as a desk and a floor, a force called **friction** exists. Friction is a force that acts in a direction opposite to the motion of the moving object. **Friction will cause a moving object to slow down and finally stop.**

Friction arises from the fact that objects and surfaces are not perfectly smooth. On a microscopic scale, the surfaces are rough. Jagged edges on one object rub against and get caught on jagged edges on the other object. Thus the amount of friction

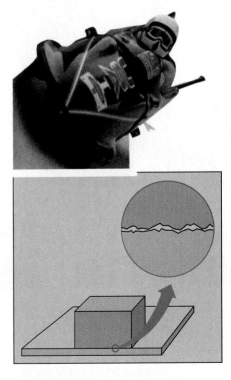

Figure 2–6 *Wheels enable roller skaters to overcome sliding friction. Yet rolling friction will cause them to slow down. How does rolling friction affect the little girls' toy?*

between two surfaces depends on how hard the surfaces are forced together and on the materials of which the surfaces are made. A heavy desk will force the surfaces together more than a light desk will. The heavier the desk you try to move, the more difficult it will be to push it across the floor. Likewise, if the floor is covered with a rough material such as carpeting, the desk will be harder to push.

The force you exert to move an object is in one direction. The force of friction is in the opposite direction. Because the two forces combine by subtraction, you must exert a force that is larger than the force of friction in order to move the object.

When solid objects slide over each other, the type of friction that results is called sliding friction. From your experience you know that sliding friction can oppose motion rather effectively. Can sliding friction be reduced? Suppose you place the object you wish to move on wheels. You can push it across the room with greater ease. You have to apply only a small amount of force because there is only a small amount of friction between the wheels and the floor. The friction produced by objects such as wheels or ball bearings is called rolling friction. Rolling friction tends to be less than sliding friction. So wheels are often placed under objects that are being moved. Just imagine how much force would have to be used if automobiles had to overcome sliding friction instead of rolling friction!

Figure 2–7 *No matter how fast a bobsled moves, it is still slowed by friction. When highly magnified, surfaces appear quite uneven and rough, making it difficult for them to slide over each other.*

Sliding friction and rolling friction describe friction between two solid surfaces. But friction also exists when an object moves across or through a fluid. All liquids and gases are fluids. Water, oil, and air are examples of fluids. The force exerted by a fluid is called fluid friction. Air resistance is an example of fluid friction caused by the particles that make up air. Air resistance makes a falling object slow down.

Fluid friction is usually less than sliding friction. Substances called lubricants, which are "slippery" substances such as grease, change sliding friction to fluid friction, thus reducing friction. Lubricants such as oil, wax, and grease are often used in devices that have moving parts, such as engines. Why?

Friction is not always a troublesome force. Friction can often be very helpful. In fact, without friction, you would not be able to walk. The friction between the soles of your shoes and the ground keeps you from slipping and sliding around. Automobiles are able to stop because the action of the brakes increases friction between the tires and the road. Cars often skid on icy streets because the smooth surface of the ice reduces the friction between the tires and the road.

Figure 2–8 *Friction is quite often helpful. For example, gymnasts use chalk on their hands to increase friction. Cyclists rely on friction to hold their bicycles on the ground during turns. And without friction, cars and other vehicles would not be able to start or stop.*

2–2 Section Review

1. What is friction? How does it affect motion?
2. Describe three types of friction.

Critical Thinking—*Making Inferences*

3. Sand is often thrown on icy walkways to prevent people from falling. Explain how the sand is helpful.

2-3 Newton's Laws of Motion

During the years 1665 and 1666, Isaac Newton developed three laws that describe all of the states of motion—rest, constant motion, and accelerated motion. In addition, these three laws explain how forces cause all of the states of motion. The importance of Newton's laws has been recognized for hundreds of years. But the significance of his contribution was perhaps best expressed by the *Apollo* crew as they were hurtling toward the moon. They radioed a message to mission control saying: "We would like to thank the person who made this trip possible . . . Sir Isaac Newton!"

Newton's First Law of Motion

Have you ever coasted on your bike along a level street? If so, you know that you continue to move for a while even though you have stopped pedaling. But do you keep on moving forever? Your experience tells you the answer to this question is no. You finally come to a stop because you are no longer exerting a force by pedaling.

The early Greek philosophers made similar observations about objects in motion. It seemed to them that in order to set an object in motion, a force had to be exerted on the object. And if that force was removed, the object would come to rest. They logically concluded that the natural state of an object was that of rest. From your everyday experiences, you would probably agree. A ball rolled along the ground comes to rest. A book pushed along a table stops sliding. A sled gliding on the snow soon stops.

But the Greeks (and perhaps you) were wrong! What brings an object to rest is friction. If there was no friction, an object would continue to travel forever. The force exerted on an object to keep it moving is simply to overcome friction. Perhaps this idea is difficult for you to imagine. After all, friction is always present in your everyday experiences. Isaac Newton, however, recognized that if friction was not present, an object in motion would continue to

Figure 2–9 *Although they would probably enjoy it, these kids will not keep moving forever. The friction between the sled and the snow will slow their movement and eventually bring them to a stop.*

Figure 2–10 *More than just going for a stroll on the cold Alaskan terrain, these sled dogs are displaying important physical properties. Sled dogs join together to exert a force great enough to overcome the inertia of the sled. What would happen if the team stopped or started suddenly?*

move forever. And an object at rest would stay at rest unless it was acted upon by an unbalanced force. You would probably agree that a football lying on a field will not suddenly fly off by itself. It will move only when thrown or kicked.

Newton called this tendency of objects to remain in motion or stay at rest **inertia** (ihn-ER-shuh). Inertia is the property of matter that tends to resist any change in motion. The word inertia comes from the Latin word *iners*, which means "idle" or "lazy." Why do you think Newton used this word? The more massive an object is, the more difficult it is to change its motion. This means that the more massive an object is, the more inertia it has. Thus the inertia of an object is related to its mass.

The concept of inertia forms the basis for Newton's first law of motion. **The first law of motion states that an object at rest will remain at rest and an object in motion will remain in motion at constant velocity unless acted upon by an unbalanced force.** Remember, constant velocity means the same speed and the same direction. In order for an object to change its velocity, or accelerate, a force must act on it. Thus, Newton's first law tells us that acceleration and force are related. There is acceleration only in the presence of forces. Any time you observe acceleration, you know that there is a force at work.

You feel the effects of inertia every day. When you are riding in a car and it stops suddenly, you keep moving forward. If you did not have a safety

Figure 2–11 *The pitch is hit and the batter strains as he begins his sprint to first base. A runner must exert more energy to start running from a stopped position than to continue running once he has begun. Why?*

belt on to stop you, your inertia could send you through the windshield. Perhaps you never thought about it this way, but safety belts protect passengers from the effects of inertia.

When you are standing on a bus you experience inertia in two ways. When the bus starts to move forward, what happens to you? You are thrown off balance and fall backward. Your body has inertia. It is at rest and tends to stay at rest, even though the bus is moving. When the moving bus stops, you fall forward. Even though the bus stops, you do not. You are an object in motion.

Because of inertia, a car traveling along a road will tend to move in a straight line. What happens, then, if the road curves? The driver turns the steering wheel and the car moves along the curve. But the people in the car continue to move in a straight line. As a result, they bump into the walls of the car. The force exerted on the people by the walls of the car keeps the people in the curved path.

Newton's Second Law of Motion

Newton's first law of motion tells you that acceleration and force are related: Acceleration cannot occur without a force. Newton's second law of motion explains how force and acceleration are related. Have you ever pushed a shopping cart along the aisles in a grocery store? If you push on the cart, it begins to move. The harder you push, the faster the cart accelerates. Thus, the greater the force, the more the acceleration. If the cart is filled with groceries, you have to push harder than you do when it is empty. This is because the cart filled with

ACTIVITY DOING

Demonstrating Inertia

Obtain a playing card or index card, several coins of different sizes, and an empty glass. Place the card on top of the glass.

Use the coins to design an experiment whose results can be explained using Newton's first law of motion. You should show that an object with more mass has more inertia.

Figure 2–12 *It is easy to make a tennis ball move with great speed after it makes contact with a racket. Although the ball has a small mass, its acceleration off the racket is great. Why?*

Figure 2–13 *Crush. Crinkle. Crash. Looks easy, but it's not. Huge trucks must exert a greater amount of force to accelerate than a small vehicle must. How does Newton's second law of motion explain this?*

ACTIVITY

CALCULATING

Move That Barge

Tugboat A exerts a force of 4000 N on a barge. Tugboat B exerts a force of 8000 N on the barge in the same direction. What is the combined force on the barge? Using arrows, draw the individual and combined forces acting on the barge. Then draw the forces involved if the tugboats were pulling in opposite directions.

groceries has more mass, or inertia. A greater force is required to accelerate an object with greater inertia. Thus force and acceleration must be related to an object's mass.

Newton's second law of motion shows how force, mass, and acceleration are related.

Force = Mass x Acceleration

When mass is in kilograms and acceleration is in meters/second/second, force is in **newtons** (N). One newton equals the force required to accelerate one kilogram of mass at one meter/second/second.

1 N = 1 kg x 1 m/sec/sec

Newton's second law of motion explains one reason why a small car has better gas mileage than a large car. Suppose the acceleration of both cars is 2 m/sec/sec. The mass of the small car is 750 kg. The mass of the large car is 1000 kg. According to the second law of motion, the force required to accelerate the small car is 750 kg x 2 m/sec/sec, or 1500 N. The force required to accelerate the large car is 1000 kg x 2 m/sec/sec, or 2000 N. More gasoline will have to be burned in the engine of the large car to produce the additional force.

Sample Problem	How much force is needed to accelerate a 1400-kilogram car 2 meters/second/second?
Solution	
Step 1 Write the formula.	**Force = Mass x Acceleration**
Step 2 Fill in given numbers and units.	**Force = 1400 kilograms x 2 meters/second/second**
Step 3 Solve for the unknown.	**Force = 2800 kilogram-meters/second/second (kg-m/sec/sec) or 2800 N**
Practice Problems	1. How much force is needed to accelerate a 66-kg skier 1 m/sec/sec?
	2. What is the force on a 1000-kg elevator that is falling freely at 9.8 m/sec/sec?

Figure 2–14 *Which of Newton's three laws of motion explains why the jumper lands in the water, not on the dock?*

Newton's Third Law of Motion

Suppose you are an astronaut making a spacewalk outside the Space Shuttle. In your excitement about your walk, you use up all of the gas in your reaction jet. How do you get back to the Shuttle?

In order to save yourself, you need to know Newton's third law of motion. **The third law of motion states that for every action, there is an equal and opposite reaction.** Another way to state the third law is to say that every force must have an equal and opposite force. All forces come in pairs.

Now, back to your problem of being stranded in space. You have no walls or floor to push against. So you throw your jet pack in the opposite direction of the Shuttle. In throwing the jet pack, you push on it and it pushes on you. The jet pack moves away from the Shuttle. You move toward safety!

You probably associate forces with active objects such as humans, animals, and machines. So it may be difficult for you to imagine that an object such as a wall or floor exerts a force. But indeed it does. This is because every material is somewhat elastic. You know that a stretched rubber band can be pulled back in such a way that it can propel a wad of paper across the room. Although other materials do not stretch as easily as rubber, they do stretch somewhat (even if you cannot see it) when a force is applied to them. And just as a stretched rubber band exerts a force to return to its original condition, so do these materials.

Take a minute now to prove this fact for yourself. Push down on the edge of your desk with your hand. The desk may not move, but your hand will have a mark on it. This mark is evidence that a

Newton's Third Law of Motion

1. Obtain a rigid cardboard strip, about 15 cm x 75 cm, a skateboard, and a motorized or windup toy car.

2. Position the skateboard upside down on the floor Place the cardboard strip and the car on top. The cardboard is the road for the car.

3. Start the car and observe what happens.

Does the car or the road move?

■ Why don't you see the road moving away from you when you are in a real car?

■ Would you be able to drive forward if you were not attached to the Earth?

Figure 2–15 *Flying gracefully through the air, birds depend on Newton's third law of motion. Could a bird fly if there was no air?*

Activity Bank

Smooth Sailing, p.140

force is being exerted on your hand. The desk is exerting the force. The harder you push on the desk, the harder the desk pushes back on your hand.

One of the most basic examples of Newton's third law of motion is walking. As you walk, your feet push against the floor. At the same time, the floor pushes with an equal but opposite force against your feet. You move forward. If the floor is highly polished, you cannot push against it with much force. So the force it exerts against your feet is also less. You move more slowly, or perhaps not at all. If you were suspended a few meters above the Earth, could you walk forward? The flight of a bird can also be explained using Newton's third law of motion. The bird exerts a force on the air. The air pushing back on the bird's wings propels the bird forward.

The reaction engine of a rocket is another application of the third law of motion. Various fuels are burned in the engine, producing hot gases. The hot gases push against the inside tube of the rocket and escape out the bottom of the tube. As the gases move downward, the rocket moves in the opposite direction, or upward.

Have you noticed that many of these examples could have been used in Chapter 1 to describe how momentum is conserved? Well, it is no coincidence. In fact, Newton arrived at his third law of motion by

Figure 2–16 *How does the third law of motion explain the movement of a water sprinkler?*

Movement of water

Movement of sprinkler arm

studying the momentum of bodies before and after collisions. The two laws are actually different ways of describing the same interactions.

Newton's three laws of motion can explain all aspects of an object's motion. His first law explains that forces are necessary to change the motion of an object. His second law describes how force and acceleration are related to mass, or inertia. His third law explains that forces act in pairs.

2–3 Section Review

1. What is inertia? How is it involved in Newton's first law of motion?
2. What three quantities are related in Newton's second law of motion? What is the relationship among them?
3. What does Newton's third law of motion say about action–reaction forces?

Connection—*You and Your World*
4. A person wearing a cast on one leg becomes more tired than usual by the end of each day. Explain this on the basis of Newton's first and second laws of motion.

2–4 Gravity

Legend has it that in the late 1500s, the famous Italian scientist Galileo dropped two cannonballs at exactly the same time from the top of the Leaning Tower of Pisa in Italy. One cannonball had ten times the mass of the other cannonball. According to the scientific theories of that day, the more massive ball should have landed first. But Galileo wanted to prove that this was not correct. He believed that the cannonballs would land at the same time. What would have been your hypothesis? According to the legend, Galileo proved to be right: Both cannonballs did land at exactly the same time! Galileo's experiment displays the basic laws of nature that govern the motion of falling objects.

Guide for Reading

Focus on these questions as you read.

▶ How is gravity related to motion?
▶ How is weight different from mass?

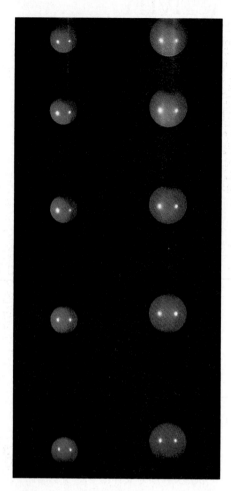

Figure 2–17 *Two objects will fall to the Earth at exactly the same rate, regardless of their masses.*

At the Center of the Gravity Matter, p.141

Falling Objects

What was so important about Galileo's discovery that a heavy object and a lighter object would land at the same time? To Isaac Newton, it meant that both objects were speeding up at the same rate, regardless of their masses. In other words, all falling objects accelerate at the same rate. A marble, a rock, and a huge boulder dropped from the top of a building at the same moment will all hit the ground at exactly the same time! According to Newton's laws of motion, if an object is accelerating, a force must be present. This force is called gravity. **The acceleration of a falling object is due to the force of gravity between the object and the Earth.**

Near the surface of the Earth, the acceleration due to the force of **gravity** (which is abbreviated as g) is 9.8 meters per second per second, or 9.8 m/sec/sec. This means that for every second an object is falling, its velocity is increasing by 9.8 m/sec. Here is an example. Suppose an object is dropped from the top of a mountain. Its starting velocity is 0 m/sec. At the end of the first second of fall, the object has a velocity of 9.8 m/sec. After two seconds, its velocity is 19.6 m/sec (9.8 m/sec + 9.8 m/sec).

Figure 2–18 *Without the force of gravity, these sky-diving acrobats would simply float in the sky. Thanks to gravity, however, they receive a thrilling adventure as they fall to Earth.*

After three seconds, 29.4 m/sec (9.8 m/sec + 9.8 m/sec + 9.8 m/sec). If it takes five seconds for the object to reach the ground, how fast will it be traveling? Perhaps you can now understand why even a dime can cause damage if it is dropped from a great height!

Activity Bank

Light Rock, p.142

PROBLEM ??? Solving

All in a Day's Work

On a colorful autumn day you head out on your newspaper route. As you are about to leave, your mother reminds you to take out the garbage. So you pick up the bag, place it on top of the newspapers that fill your wagon, and drop it off at the curb. On your way along the sidewalk, you come across the neighbor's cat, which you lift up and briefly pet before it jumps down and runs off. You continue on your way, pulling your wagon filled with newspapers behind you. At the next house, you strategically throw the paper from the sidewalk to the front step. Perfect shot! As you head for your next stop, you see a few acorns hanging from a tree. You pull them off the tree and throw them on the sidewalk ahead of you. Shortly after that, you see two of your friends trying to push a heavy bag of leaves. They are not having much success, so you join in and the three of you move the bag to the side of the house. You say goodbye to your friends and continue on your way until your red wagon is totally empty and it's time to go home—just in time for dinner.

Making a Diagram

During this short walk, a number of forces were exerted. Draw a series of diagrams showing each activity you performed. Use stick figures and arrows to show the forces involved. Do not forget about friction and gravity! When you finish, think of some other activities that might have taken place during your walk: catching a ball, moving a branch, lifting a rock. Add these to your drawings.

ACTIVITY

DISCOVERING

Science and Skydiving

1. Design a parachute using a large bandana and a piece of string or thread. Attach a clothespin to it.

2. Drop the parachute and a second clothespin from the same height at the same time. **CAUTION:** *Do not climb on any object without adult approval and supervision.* Which do you expect to hit the ground first? Why? Are you correct?

3. Describe the motion of the objects as they fell.

■ Redesign your parachute to decrease its velocity.

■ Can you now explain why insects can fall from tremendous heights yet walk away unharmed? (*Hint:* Compare their masses to their surface areas.)

Figure 2–19 *Although gravity pulls both a leaf and a rock toward the Earth, the two objects do not accelerate at the same rate on the Earth. Astronauts, however, have found that the two objects land at the same time on the moon. Why?*

Air Resistance

Do a leaf, a piece of paper, and a feather fall at 9.8 m/sec/sec? Because you have probably seen these objects fluttering through the air to the ground, you know the answer is no. Their acceleration is much less than 9.8 m/sec/sec. Why? As a leaf falls, air resistance opposes its downward motion. So it moves more slowly. Air resistance also opposes the downward motion of a falling rock. But the shape of the leaf causes greater air resistance, and so its downward motion is more significantly slowed. If both the leaf and the rock were dropped in a vacuum, they would accelerate at 9.8 m/sec/sec.

Any falling object meets air resistance. You can think of the object as being pushed up by this opposing force of the air. As the object falls, the air resistance gradually becomes equal to the pull of gravity. The forces are then balanced. According to the first law of motion, when forces are balanced there is no acceleration. The object continues to fall, but it falls at a constant velocity. There is no further acceleration. When a falling body no longer accelerates (but continues to fall at a constant velocity), it has reached its terminal (final) velocity. Sky divers cannot accelerate any further once they reach a terminal velocity of about 190 km/hr. At this point, the sky divers continue their descent, although there is no longer any sensation of falling!

Newton's Law of Universal Gravitation

Although his work had provided answers to so many questions about falling objects and gravity, Newton did not stop there. He went even further. He wondered if the force that was making the apple fall to the Earth was the same force that kept the moon in its path around the Earth. After all, since the direction of the moon was constantly changing in its circular path, it too was accelerating. Therefore a force must be involved.

Newton calculated the acceleration of the apple and compared it with the acceleration of the moon. Using laws already presented by Johannes Kepler (1571–1630), a brilliant astronomer, Newton was able to derive a formula to calculate the force acting on both the apple and the moon. He concluded that the force acting on the moon was the same force that was acting on the apple—gravity.

In Newton's day, most scientists believed that forces on the Earth were different from forces elsewhere in the universe. Newton's discovery represented

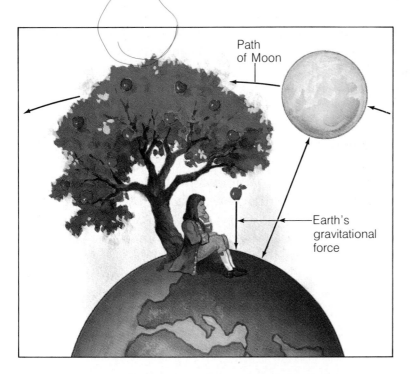

Path of Moon

Earth's gravitational force

Figure 2–20 *Newton's universal law of gravitation explains why an apple falls to the ground as well as why the moon stays in its orbit around the Earth.*

ACTIVITY DOING

Science and the Leaky Faucet

1. Adjust a faucet so that it slowly drips. Use a deep sink, if possible.

2. Measure the distance from the tip of a hanging drop to the bottom of the sink.

3. Use a stopwatch to measure the time it takes for the drop to fall to the bottom of the sink.

4. Calculate the average velocity of the drop. Repeat steps 3 and 4 three more times.

5. Use your average value of velocity to calculate the acceleration of the water drop. This is the acceleration due to gravity.

How close does your value come to the accepted value of 9.8 m/sec/sec? If your value does not match the accepted value, what reasons can you give for the difference?

Activity Bank

Putting Gravity to Work, p.143

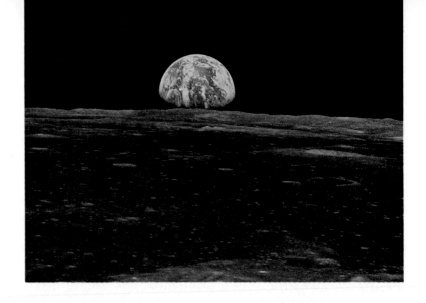

Figure 2–21 *Although the Earth's gravitational attraction decreases as distance increases, a force is still exerted at a distance as great as that of the moon. In fact, the Earth's gravitational attraction is responsible for holding the moon in its path. How is gravity related to the organization of the solar system?*

the first universal law of forces. A universal law applies to all objects in the universe. Newton's **law of universal gravitation** states that all objects in the universe attract each other by the force of gravity. The size of the force depends on two factors: the masses of the objects and the distance between them.

The force of gravity increases as the masses of the objects increase. Although gravitational forces always exist between objects, they only become observable when the masses of the objects are as large as those of the planets, moon, and stars. For example, there is a force of gravity between you and this textbook. Yet the textbook is not pulled over to you. Why? The force of gravity depends on the masses of the objects. The gravitational force between a book and you is extremely small because your mass and the book's mass are small compared with the mass of the Earth.

Gravitational force decreases rapidly as the distance between objects increases. The gravitational force between an apple and the Earth is about 2 N on the surface of the Earth. At 380,000 km—the distance to the moon—the gravitational force between the apple and the Earth is only 0.001 N.

Gravity is of great importance in the interactions of large objects. It is gravity that binds us to the Earth and holds the Earth and other planets in the solar system. The force of gravity plays an important role in the evolution of stars and in the behavior of galaxies. In a sense, it is gravity that holds the universe together.

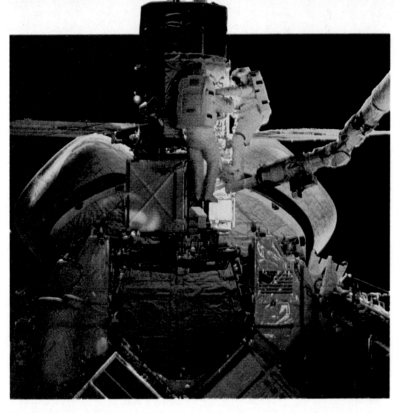

Figure 2-22 *Making repairs or performing experiments is quite a different experience in an almost weightless environment. For example, grabbing a huge piece of machinery in only one hand would be a great advantage. However, the fact that forgetting to tie one's self onto the work station would result in floating up and away would not be as favorable.*

ACTIVITY READING

Voyage to the Moon

Before it was possible to travel to the moon, Jules Verne envisioned such a trip. In Verne's *From the Earth to the Moon,* he described how people would get to the moon and how they would have to adjust to the conditions outside the Earth's atmosphere. Read the book and discover situations Verne's space travelers encounter.

Weight and Mass

You are all familiar with the term weight. Each time you step on a scale, you are looking to see what you weigh. **Weight is a measure of the force of gravity on an object.** In this case, the object is you! Since weight is a force, its unit is the newton (N). This textbook weighs about 15 N. A medium-sized car probably weighs between 7000 and 9500 N.

Your weight varies according to the force of gravity pulling on you. And the force of gravity varies according to distance from the center of the Earth. Suppose you weigh yourself in the morning at sea level. Later that day you ride to the top of a tall mountain, weigh yourself again, and find that you weigh less. What has happened? Is there less of you on top of the mountain than there was at sea level? After all, you weigh less. The answer, of course, is no. In a given day, you have the same amount of

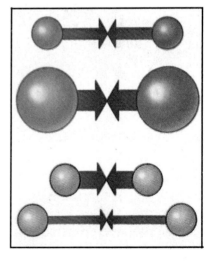

Figure 2–23 *The force of attraction between two objects increases as mass increases (top). It decreases as distance increases (bottom). The wider the arrow, the greater the force.*

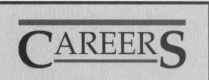

mass regardless of your location. Your mass does not change, unless of course you diet and exercise. Your mass is the same anywhere on the Earth (on top of a mountain or at sea level), on the moon, and even on Jupiter. It is your weight that changes. Because you are farther from the center of the Earth when you are on top of a mountain than when you are at sea level, the pull of gravity on you decreases. Thus you weigh less on top of the mountain than you do at sea level.

Although mass and weight are not the same thing, they are related. This may be obvious to you because you know that more massive objects weigh more than less massive objects. Newton's second law of motion, force = mass x acceleration, can be rewritten in terms of weight to show the relationship.

Weight = Mass x Acceleration due to gravity

$$w = m \times g$$

Remember that the unit of weight is the newton and the unit of mass is the kilogram.

On the surface of the Earth, the acceleration due to gravity is 9.8 m/sec/sec. A 10-kg mass would weigh 10 kg x 9.8 m/sec/sec, or 98 N. If your mass is 50 kg, your weight would be 490 N. What would be the weight of a 100-kg mass?

2–4 Section Review

1. How is gravity related to falling objects?
2. How would all objects accelerate if they fell in a vacuum? Why?
3. What does the law of universal gravitation state?
4. Compare weight and mass.

Critical Thinking—*Making Calculations*
5. An astronaut who weighs 600 N on Earth is standing on an asteroid. The gravitational force of the asteroid is one hundredth of that of the Earth. What is the astronaut's weight on the asteroid?

Which Way Is Up?

You know that the sky is up and the ground is down because you can see it. But would you know the same if you closed your eyes? Yes, you would! Astronauts in space can spin comfortably in all directions, but you cannot. Even with your eyes closed, you can tell which way is up and which way is down. You can even determine if you are moving and in what direction. You have your *ears* to thank for all this!

In one area of the inner ear, special structures called otoliths determine whether the body is speeding up, slowing down, or changing direction. They do this by comparing the body's movement with something that is always in the same direction—the downward force of gravity. When the otoliths move, they pull on hair cells that relay a nerve impulse to the brain describing the position or motion of the head. For example, when your head is in the upright position, gravity pulls the otoliths down. The otoliths in turn push the sensory hairs down, rather than to one side or the other. When your head is tilted, the pull of gravity shifts the otoliths to the side. This causes the sensory hairs to send a different signal to the brain.

Additional balance comes from another section of the inner ear. Here three tiny canals called semicircular canals lie at right angles to each other. A fluid flows through each canal in response to motion in a particular direction. If, for example, you move your head from right to left when you say "no," the fluid in the semicircular canal that detects horizontal motion will be forced to move. When the fluid moves, it disturbs hair cells that send messages about the movement to the brain. Because each semicircular canal detects motion in one dimension, the arrangement of the canals enables a person to detect movement in all directions.

So the next time you find yourself upside down, give some thought to what your ears have to do with it!

The otoliths (left) and the semicircular canals (right) enable a gymnast to maintain balance as she completes a back flip.

Laboratory Investigation

Will an Elephant Fall Faster Than a Mouse?

Problem

Does mass affect the rate of fall?

Materials *(per group)*

wood block, 10 cm x 15 cm x 2.5 cm
Styrofoam pad, 10 cm x 15 cm x 2.5 cm
sheet of notebook paper
triple-beam balance

Procedure

1. Use the triple-beam balance to determine the masses of the block, Styrofoam pad, and paper. Record each mass to the nearest 0.1 gram.

2. Hold the block and foam pad horizontally at arm's length. The largest surface area of each object should be parallel to the ground.

3. Release both the block and the foam pad at the same time. Observe if they land at the same time or if one hits the ground before the other.

4. Repeat step 3 several times. Record your results.

5. Repeat steps 2 to 4 for the foam pad and the paper.

6. Crumple the paper into a tight ball.

7. Compare the falling rates of the crumpled paper and the foam pad. Record your observations.

8. Compare the falling rates of the crumpled paper and the wood block. Record your observations.

Observations

Object	Mass	Falling Rate (comparative)
Wood block		
Styrofoam pad		
Paper (uncrumpled)		
Paper (crumpled)		

1. Which reaches the ground first, the wood block or the foam pad?

2. Are your results the same in each trial?

3. Which reaches the ground first, the foam pad or the uncrumpled paper?

4. Which reaches the ground first, the foam pad or the crumpled paper?

Analysis and Conclusions

1. Galileo stated that two bodies with different masses fall at the same rate. Do your observations verify his hypothesis? Explain your answer.

2. Did crumpling the paper have any effect on its falling rate? Explain your answer.

3. Now answer this question: Would an elephant fall faster than a mouse? Explain your answer.

4. **On Your Own** Design and perform an experiment that compares different objects made out of the same material.

Summarizing Key Concepts

2–1 What Is Force?

▲ A force is a push or pull. A force may give energy to an object, setting the object in motion, stopping it, or changing its direction.

▲ Forces in the same direction combine by addition. Forces in opposite directions combine by subtraction.

▲ Unbalanced forces cause a change in motion. When forces are balanced, there is no change in motion. Balanced forces are opposite in direction and equal in size.

2–2 Friction: A Force Opposing Motion

▲ Friction is a force that opposes motion.

▲ The three kinds of friction are sliding, rolling, and fluid friction.

2–3 Newton's Laws of Motion

▲ Inertia is the tendency of matter to resist a change in motion.

▲ Newton's first law of motion states that an object at rest will remain at rest and an object in motion will remain in motion at constant velocity unless acted upon by an unbalanced force.

▲ Newton's second law of motion describes how force, acceleration, and mass are related. Force equals mass times acceleration.

▲ Newton's third law of motion states that forces always occur in pairs. Every action has an equal and opposite reaction.

2–4 Gravity

▲ The acceleration due to gravity at the surface of the Earth is 9.8 m/sec/sec.

▲ Gravity is a force of attraction that exists between all objects in the universe.

▲ The size of the force of gravity depends on the masses of the two objects and the distance between them.

▲ Weight and mass are different quantities. Weight is a measure of the pull of gravity on a given mass. Mass is a measure of the amount of matter in an object. Mass is constant; weight can change.

Reviewing Key Terms

Define each term in a complete sentence.

2–1 What Is Force?
force

2–2 Friction: A Force Opposing Motion
friction

2–3 Newton's Laws of Motion
inertia
newton

2–4 Gravity
gravity
law of universal gravitation

Chapter Review

Content Review

Multiple Choice

Choose the letter of the answer that best completes each statement.

1. Force is
   a. a push.   c. the ability to change motion.
   b. a pull.   d. all of these answers

2. Forces that are opposite and equal are called
   a. balanced.   c. unbalanced.
   b. friction.   d. gravitational.

3. The force that opposes the motion of an object is called
   a. acceleration.   c. density.
   b. friction.   d. gravity.

4. The type of friction that exists for a shark swimming in the ocean is
   a. sliding.   c. rolling.
   b. hydraulic.   d. fluid.

5. The property of matter that resists a change in motion is
   a. inertia.   c. gravity.
   b. friction.   d. weight.

6. According to Newton's second law of motion, force equals mass times
   a. inertia.   c. direction.
   b. weight.   d. acceleration.

7. The force of attraction that exists between all objects in the universe is
   a. friction.   c. momentum.
   b. inertia.   d. gravity.

8. A change in the force of gravity pulling on you will change your
   a. mass.   c. inertia.
   b. air resistance.   d. weight.

True or False

If the statement is true, write "true." If it is false, change the underlined word or words to make the statement true.

1. A <u>force</u> can set an object in motion, stop its motion, or change the speed and direction of its motion.

2. The combined force of <u>unbalanced</u> forces is always zero.

3. Friction is a force that always acts in a direction <u>opposite</u> to the motion of the moving object.

4. "Slippery" substances such as oil, wax, and grease that reduce friction are called <u>lubricants</u>.

5. Objects in constant motion will remain in constant motion unless acted upon by <u>balanced</u> forces.

6. Force equals mass times <u>velocity</u>.

7. In a vacuum, a heavier object will fall to the Earth <u>faster than</u> a lighter object will.

8. <u>Mass</u> is the measure of the force of gravity.

Concept Mapping

Complete the following concept map for Section 2–3. Refer to pages S6–S7 to construct a concept map for the entire chapter.

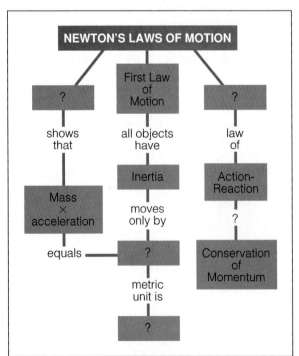

Concept Mastery

Discuss each of the following in a brief paragraph.

1. Why is there a force involved when the sails of a windmill turn?
2. Distinguish between balanced and unbalanced forces.
3. Why do athletes' shoes often have cleats on them?
4. Explain how Newton's three laws explain all aspects of an object's motion.
5. Explain why a single force cannot exist.
6. When a golf ball is dropped to the pavement, it bounces up. Is a force needed to make it bounce up? If so, what exerts the force?
7. Why does a raindrop fall to the ground at exactly the same rate as a boulder?
8. Explain why a flat sheet of paper dropped from a height of 2 meters will not accelerate at the same rate as a sheet of paper crumpled into a ball.
9. What is the relationship between weight and mass?

Critical Thinking and Problem Solving

Use the skills you have developed in this chapter to answer each of the following.

1. **Applying concepts** How is inertia responsible for removing most of the water from wet clothes in a washing machine?
2. **Making connections** Although Newton's first law of motion has two parts, they actually say the same thing. Explain how this can be true using what you learned about frames of reference.
3. **Applying concepts** Suppose a 12-N force is required to push a crate across a floor when friction is not present. In reality, friction exerts a force of 3 N. If you exert a force of 7 N, what size force must your friend exert so that you can move the crate together? Draw a diagram showing the forces involved.

4. **Making generalizations** What happens to the force of gravity if mass increases? If distance increases? Write a statement that explains how gravity, mass, and distance are related according to Newton's law of universal gravitation.
5. **Making calculations** A heavy object is dropped from the top of a cliff. What is its velocity at the end of 2 seconds? At the end of 5 seconds? Just before it hits the ground after 12 seconds?
6. **Identifying relationships** Suppose the acceleration due to gravity on a planet called Zorb is 20 m/sec/sec. What is the weight of a 100-kg Zorbian?
7. **Using the writing process** Pretend you live on a planet whose size is the same but whose mass is four times greater than the mass of the Earth. Using the words listed below, write a 200-word story about a typical day in your life. Include information about how your planet's mass affects you and the life forms around you.

acceleration   law of universal gravitation
gravity   momentum
inertia   weight

Forces in Fluids

Guide for Reading

After you read the following sections, you will be able to

3–1 Fluid Pressure
- Describe how the particles of a fluid exert pressure.

3–2 Hydraulic Devices
- Explain how a hydraulic device operates.

3–3 Pressure and Gravity
- Relate fluid pressure to altitude and depth.

3–4 Buoyancy
- Describe the relationship between the buoyant force and Archimedes' principle.

3–5 Fluids in Motion
- Explain why an object floats or sinks.
- Recognize how Bernoulli's principle is related to flight.

It is December 17, 1903. Wilbur and Orville Wright stand on a deserted beach in Kitty Hawk, North Carolina. Orville climbs into a strange-looking seat made of wood and canvas. A 12-horsepower gasoline engine is connected to two large propellers by a chain and sprocket. The Wright brothers are about to try something no one has ever succeeded in doing before. They are going to fly this machine!

They have prepared well for their attempt at flight. For the past 25 years they have studied the dynamics of air flight. They have experimented with more than 200 different wing surfaces in their homemade wind tunnel. They have observed and analyzed the flight of buzzards, carefully noting how the birds turn in the sky without losing balance.

Now they are finally ready. In a flight that lasts just 12 seconds, the plane manages to travel 36 meters. It is a small distance, but a significant step in science: Human flight has become a reality!

The first flying machine was designed in the fifteenth century by Leonardo da Vinci. Why did it take so long to fly the first plane? How can a jumbo jet fly over 800 kilometers per hour? As you read this chapter, you will learn the answers.

Journal *Activity*

You and Your World Can you remember a time when you got caught in a storm? Did the wind pushing against you almost stop you in your tracks? In your journal, describe the storm and how it felt to get caught in it. Include details such as the puddle you stepped in or the snow that blew into your coat or gloves.

◀ *As Wilbur Wright stood watching on the deserted beach at Kitty Hawk, North Carolina, his brother Orville took one of the most important trips in history—a 12-second, 36-meter leap toward the attainment of human flight.*

3–1 Fluid Pressure

When you think of forces and Newton's laws of motion, do you think only of solid objects—pushing a box, pulling a wagon, lifting a crate? Although you may not realize it, forces exist naturally in fluids as well. Fluids are substances that do not have a rigid shape. Liquids and gases are fluids. When you breathe, when you swim, when you drink from a straw, you are experiencing forces created by fluids. As a matter of fact, there is a force approximately equal to the weight of an automobile pushing down on you right now! Do you know why?

What Is Pressure?

All matter is made up of tiny particles. **The forces that exist in fluids are caused by the mass and motion of the particles making up the fluid.** In a solid, the particles are packed very tightly together. There is very little movement of the particles in a solid. In liquids and gases, however, the particles are not packed together so tightly. Thus they are able to move about more freely. The particles that make up fluids are moving constantly in all directions. As each particle moves, it pushes against other particles and against the walls of its container with a force that depends on the mass and acceleration of the particle. The "push," or force, particles exert over a certain area is called **pressure**. Fluid pressure is exerted equally in all directions.

Perhaps you are familiar with the word pressure as it is used to describe water, air, and even blood. Scientists define pressure as force per unit area. Pressure can be calculated by dividing the force exerted by a fluid by the total area over which the force acts:

$$\textbf{Pressure} = \frac{\textbf{Force}}{\textbf{Area}}$$

Figure 3–1 *Although they may not realize it, these wind surfers and cliff divers could not enjoy their activities without the forces exerted by fluids. What fluids are involved in the actions shown in these photos?*

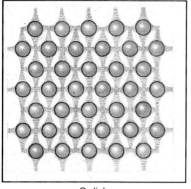
Solid

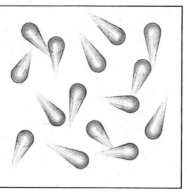
Liquid

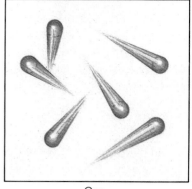
Gas

Figure 3–2 *The arrangement and movement of the particles that make up a substance determine the characteristics of the substance. Notice that as you move from solids to gases, the particles become more spread out and motion increases. How does this explain why gases exert the greatest pressure?*

When force is measured in newtons (N) and area is measured in square centimeters (cm^2), pressure is measured in newtons per square centimeter (N/cm^2).

Air in the atmosphere exerts a pressure of 10.13 N/cm^2 at sea level. If your back has an area of approximately 1000 cm^2, then you have a force of 10,130 N pushing on your back. This is the force approximately equal to the weight of an automobile you read about earlier. What keeps this force from crushing you? The fluids inside your body also exert pressure. The air pressure outside your body is balanced by the fluid pressure inside your body. So you do not feel the outside force.

You are probably familiar with some important consequences of pressure. Many devices must be inflated to a particular pressure before they can operate properly. For example, a car whose tires are not properly inflated may not ride correctly or get its expected gas mileage. What will happen if a basketball is not filled to the proper pressure? Meteorologists also pay careful attention to pressure. Atmospheric pressure is an important indicator of weather conditions. High and low pressure areas are each associated with specific weather characteristics.

Differences in Pressure

You probably did not give much thought to what you were doing the last time you drank through a straw. But what you actually do when you suck on a straw is remove most of the air from inside the straw. This causes the pressure inside the straw to

Activity Bank

Watering Your Garden Green, p.145

Figure 3–3 *This can was crushed because of a change in air pressure. Was the air pressure greater inside the can or outside it?*

Figure 3–4 *It would be very difficult for this girl to enjoy her ice cream soda if it were not for unequal air pressure. The air pressure pushing down on the liquid outside the straw is greater than the air pressure inside the straw. This difference in pressure forces the liquid up.*

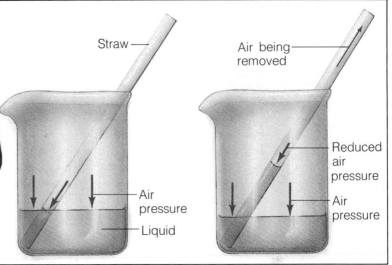

Straw

Air being removed

Air pressure

Liquid

Reduced air pressure

Air pressure

ACTIVITY
DOING

Air Pressure

1. Obtain an empty plastic 1-liter bottle that has an airtight top.

2. Fill the bottle one-quarter full with hot water.

3. Tightly secure the top to the bottle so that no air can enter or escape.

4. Place the bottle in a refrigerator for about five minutes. At the end of this time, record the shape of the bottle.

When the hot air in the bottle cools, the particles slow down and do not push into each other as often. Thus, the air pressure inside the bottle decreases.

What causes the bottle to collapse?

decrease. Standard air pressure, which is now greater than the air pressure inside the straw, pushes down on the surface of your drink. This push forces the drink up through the straw and into your mouth! See Figure 3–4. The principle that enables you to drink from a straw is an important property of fluids. **Fluids will move from areas of higher pressure to areas of lower pressure.**

The operation of a vacuum cleaner is another example of the principle of unequal air pressure. It may surprise you to learn that a vacuum cleaner does not suck up only dirt. A fan inside the cleaner causes the air pressure within the machine to become less than the pressure of the air outside the machine. The outside air pressure pushes the air and dirt into the vacuum cleaner. A filter then removes the dirt and releases the air. In addition to vacuum cleaners, all devices that involve suction take advantage of differences in pressure. This includes plungers, suction cups, and even medicine droppers.

One extremely important consequence of the principle of unequal pressure is your ability to breathe. When you breathe, you use a large muscle at the base of your rib cage to change the volume and pressure of the chest cavity. This muscle is called the diaphragm. When you inhale, the diaphragm flattens, giving your lungs room to expand. In turn, the air particles inside your lungs have more room to move about, which means that the pressure decreases. When the pressure inside your lungs is less than the air pressure outside your body, air is forced through your mouth or nose into your lungs. When you exhale, the diaphragm moves upward, reducing the size of your lungs. This causes the pressure inside your lungs to increase to a pressure greater than that outside your body. Air is now forced out.

3–1 Section Review

1. Why does fluid pressure exist?
2. How is pressure calculated?
3. Explain how a woman weighing 500 N and wearing high-heeled shoes can exert a pressure on the floor equal to about three times the pressure exerted by a 45,000-N elephant.

Connection—*Life Science*
4. What can happen to a person's blood vessels if his or her blood pressure gets too high?

Figure 3–5 *Pop. Fizz. These are familiar sounds associated with opening certain containers. Many liquids are put into cans or bottles under high pressure. When the container is opened, that pressure is released.*

3–2 Hydraulic Devices

If there are no outside forces acting on a fluid, the pressure exerted by the fluid will be the same throughout. And the pressure will be exerted in every direction—up, down, sideways. Suppose you have a balloon filled with air and you poke your finger into it without popping it. Your finger adds pressure to the air at that point inside the balloon. The particles of the air are already packed tightly together and cannot escape. So what happens to this additional pressure applied to the air? The pressure at any point in a fluid is transmitted, or sent out, equally in all directions throughout the fluid. This means that the pressure is increased in every direction.

You have probably experienced this event without even knowing it. If you have a bottle completely filled with water and you try to push a stopper into it, what happens? You probably get wet as the water squirts out the top. The pressure applied to the water by

Guide for Reading

Focus on this question as you read.

▶ *How do the properties of fluid pressure make hydraulic devices operate?*

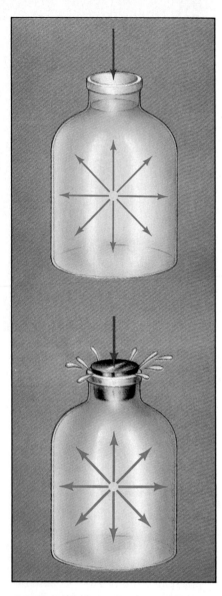

Figure 3–6 *A liquid in a confined space such as a bottle exerts pressure equally in all directions. When a stopper is pushed into the bottle, the added pressure it exerts is also transmitted equally in all directions—including up.*

the stopper acts equally in all directions—including up! **The transmission of pressure equally in all directions in a liquid is the principle behind hydraulic devices.** The brakes on your family car and a hydraulic lift used to raise heavy objects are examples of **hydraulic devices.** Hydraulic devices produce enormous forces with the application of only a very small force. In other words, hydraulic devices multiply forces. Let's see just how this works in the case of hydraulic brakes.

You may have wondered how it is possible that a rapidly moving car with a mass of more than 1000 kilograms can be stopped with a relatively light push on the brake pedal—a push certainly much lighter than you would need to exert if you were trying to stop the car from the outside. Imagine two movable pistons connected to a container of liquid as shown in Figure 3–7. The smaller piston can be pushed downward. This piston is like the piston connected to the brake pedal. When a force is exerted on the piston, the pressure created by the force pushes against the liquid. The pressure is transmitted equally throughout the liquid.

Now for the surprising part. The force experienced by the larger piston is greater than the force used to move the smaller piston. How does

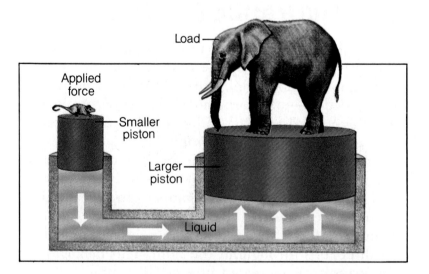

Figure 3–7 *In a hydraulic lift, the applied force moves the smaller piston down and adds pressure to the liquid. That pressure is transmitted equally in all directions, so the same pressure is exerted on the larger piston. But because the area of the larger piston is greater than the area of the smaller piston, more force is produced. The larger piston and its load move up.*

this happen? The pressure on every square centimeter of the larger piston is equal to the pressure on every square centimeter of the smaller piston. The force exerted on a piston is the pressure of the liquid times the area of the piston. Since the number of square centimeters (or area) is greater on the larger piston, the force is greater. In the case of hydraulic brakes, the larger piston would be connected to the brake pads that slow the tires down. This is how a push on the brake pedal can bring a car to a halt.

You may already be familiar with some hydraulic devices. Barbers' chairs, automobile lifts, rescue ladders, and robotic equipment all use such devices. In addition, a number of living organisms make use of hydraulic pressure. A sea anemone can achieve a variety of shapes by the action of muscles on its seawater-filled body cavity. Earthworms move forward by repeated contractions of circular muscles along the body that act on their fluid-filled body cavities. The legs of some spiders are not extended only by muscles. Instead, the legs are extended by fluid driven into them under pressure.

Figure 3–8 *Besides being practical, hydraulic devices sometimes sustain life, as for this sea anemone, or provide endless fun, as for these children at the amusement park. No matter what the application, hydraulic devices are used to multiply force.*

3–2 Section Review

1. Explain the principle used in the operation of hydraulic devices.
2. Name four hydraulic devices with which you are familiar.

Critical Thinking—*Applying Concepts*

3. Without changing the size of the applied force or the small piston, how can you increase the amount of force that comes out of a hydraulic lift or ladder?

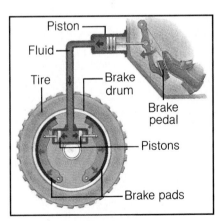

Figure 3–9 *Fluid in the brake system of a car multiplies the force exerted by the driver into a force great enough to stop the moving car. What force is used by the brake pads to stop the tire from turning?*

3–3 Pressure and Gravity

You learned that the pressure exerted by a fluid is the same throughout the fluid if there are no forces acting on the fluid. But there is one force that is always present. That force is gravity. Gravity pulls downward on all of the particles in a fluid.

The force of gravity produces some familiar results. For example, if you have ever swum to the bottom of a pool, then you remember how your ears began to ache as you went deeper. This happened because the pressure of the water increased rapidly with depth. **Due to the force of gravity, the pressure of any fluid varies with its depth.** The greater the depth, the greater the pressure. Let's see why.

The pool of water in Figure 3–10 has been broken up into five different levels. Because gravity pulls down on the particles in the top level, the entire level has a certain weight. The force of the weight of the first level pushes down on the second level. The second level, then, has the pull of gravity on its own particles plus the force of the weight of the first level. Therefore, the pressure at the second level is greater than the pressure at the first level. What about at the third level? The third level has the pull of gravity on its own particles plus the weight of the first two levels pushing down on it. So the pressure at the third level is greater than at either level above it. The bottom level (or greatest depth) of any fluid will have the greatest pressure because it has the greatest force pushing down on it from all the levels above it.

The increase in pressure that accompanies an increase in depth has some important effects.

Figure 3–10 *As any diver knows, the pressure in a fluid increases with depth. Where is the pressure greatest in a swimming pool?*

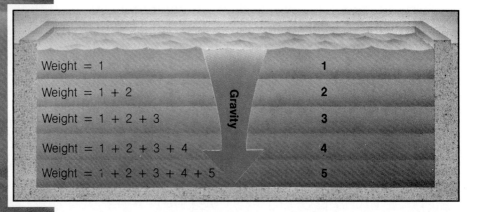

Weight = 1	1
Weight = 1 + 2	2
Weight = 1 + 2 + 3	3
Weight = 1 + 2 + 3 + 4	4
Weight = 1 + 2 + 3 + 4 + 5	5

Gravity

Figure 3–11 *Because pressure increases downward, the stream from the bottom hole in this container of water is strongest. Try it and see! The horizontal supporting ribs of a silo are closer together at the bottom than at the top. Why?*

Submarines that have descended too deep in the ocean have on occasion been crushed by the tremendous pressure. Divers cannot go too deep without experiencing serious problems caused by the increased pressure. Under high-pressure conditions, more nitrogen gas than usual dissolves in a diver's blood. When the diver resurfaces, the pressure on the body greatly decreases and the nitrogen gas leaves the blood. If the nitrogen leaves too quickly, it forms tiny bubbles that are often quite painful and can be dangerous. This condition is sometimes referred to as the bends. In order to prevent the bends, a diver must rise slowly to allow the dissolved nitrogen to be released gradually from the blood.

Water is not the only fluid in which pressure varies with depth. Our planet is surrounded by a fluid atmosphere. The pressure of our atmosphere also varies. In this case, the pressure varies with altitude, or height above the ground. The higher the altitude, the lower the pressure. In addition, at higher altitudes, there are fewer particles of air in a given area. Fewer particles pushing against one another results in lower pressure. At higher altitudes, the pressure inside your body becomes greater than the air pressure outside your body. You may feel the difference in pressure as a pain in your eardrums. When this happens, some air rushes out of your ears and you hear a "pop." As a result of the release of some of the air from inside your eardrum, the pressure inside your eardrum is again equal to the pressure outside your body.

ACTIVITY

DISCOVERING

A Plumber's Magic

1. Wet the bottom rim of a plumber's plunger.

2. Push the plunger tight against the seat of a stool, the chalkboard, or some other smooth surface.

What happens when you try to lift the plunger?

■ Using what you know about air pressure, explain how a plunger works.

3-3 Section Review

1. What force causes fluid pressure to vary with depth? Why?
2. Why must divers rise slowly?
3. Why are dams thicker at the bottom than at the top?

Connection—*Life Science*

4. Explain why it is more difficult to breathe at higher altitudes than at sea level.

Guide for Reading

Focus on these questions as you read.

▶ *What is buoyancy?*
▶ *How does Archimedes' principle explain why an object floats or sinks?*

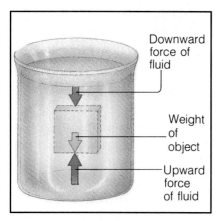

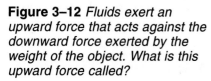

Figure 3–12 *Fluids exert an upward force that acts against the downward force exerted by the weight of the object. What is this upward force called?*

3-4 Buoyancy

Here is an experience that you have probably had. You have been able to lift a friend or heavy object while you were in or under water that you could not lift while you were out of the water. Objects submerged in a fluid appear to weigh less than they do out of the fluid. Why? Force (pressure) increases with depth. Thus the force at the bottom of an object in a fluid is greater than the force at the top of the object. The overall force is in the upward direction and acts against the downward weight of the object. The upward force is called the **buoyant** (BOI-uhnt) **force.** This phenomenon is known as **buoyancy.**

Think for a moment about what happens to the level of water in a bathtub when you sit down in it. The level rises. It does so because you move aside some of the water and take its place. Any object placed in water displaces, or moves aside, a certain amount of water. The amount of water that is displaced has a definite weight. Because the buoyant force was able to support this weight, this weight must be related to the size of the buoyant force.

More than 2000 years ago, the Greek scientist Archimedes discovered the nature of this relationship. **The buoyant force on an object is equal to the weight of the fluid displaced by the object.** This relationship between buoyancy and the weight of the displaced fluid is called **Archimedes' principle.**

The size of the buoyant force determines what will happen to an object placed in a fluid—that is, whether it will sink or float. The buoyant force can be greater than, less than, or equal to the weight of an object placed in the fluid. What do you think happens when the weight of an object placed in a fluid is less than or equal to the weight of the fluid it displaces—and therefore the buoyant force? You are correct if you said the object floats. **An object floats when it displaces a volume of fluid whose weight is greater than or equal to its own weight.**

Have you ever heard the expression "tip of the iceberg" applied to a situation in which only a few facts are known (the rest are hidden)? This expression is based in science—in particular, in Archimedes' principle. An iceberg is a massive chunk of ice that has broken away from a glacier and is floating in the ocean. Because it is floating, you know that its weight must be less than or equal to the weight of the salt water it displaces. A volume of ice weighs slightly less than the same volume of salt water. Therefore, the buoyant force of the ocean pushes the iceberg upward. But because the weight of the iceberg and the weight of the displaced water are so close, the iceberg is pushed upward only a small amount. Close to 90 percent of the iceberg remains submerged. Icebergs can be extremely dangerous because a passing ship may see only the small portion that is above water—the tip—and thus be damaged by the much larger portion floating beneath the surface.

Exactly why do some substances float and others sink? The answer has to do with a physical property of both the object and the fluid called **density.** Density is the ratio of the mass of a substance to its volume. In other words, density is mass divided by volume ($D = M/V$). Here is an example. A block of wood placed in water will float. But the same size block of aluminum placed in water sinks. Why? For the displaced water to have a greater weight than the object (the condition for floating), the fluid must have a greater density than the object. Water, then, must be more dense than wood, but less dense than aluminum.

The conditions for floating can now be stated in terms of density: **An object will float in a fluid if the density of that object is less than the density of the fluid.** The density of water is 1 gram per cubic

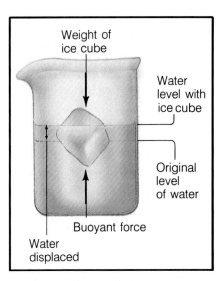

Figure 3–13 *An object placed in a fluid displaces an amount of fluid causing the level to rise. The buoyant force is equal to the weight of the displaced fluid. How can you measure the amount of water displaced by this ice cube?*

Activity Bank

Density Dazzlers, p.147

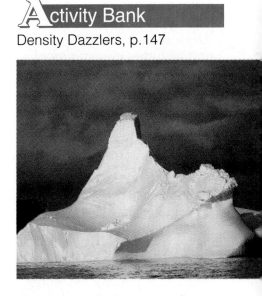

Figure 3–14 *Do not worry about the tip of an iceberg. It is what lies beneath the surface that is the problem! Because the density of ice is slightly less than the density of water, only a portion of an iceberg needs to be pushed up through the surface for the iceberg to float.*

ACTIVITY

An Archimedean Trick

1. Obtain a cubic object such as a die.

2. Measure the volume of the cube. Volume equals length times width times height.

3. Obtain a graduated cylinder or small beaker into which you can place the cube. Fill half of the container with water. Record the volume of the water either by reading the level on the graduated cylinder or making a mark on the beaker with a glass-marking pencil.

4. Place the cube into the water. Make sure the cube is entirely under water. Record the volume shown on the container again.

5. Subtract the first volume measurement from the second volume measurement. Compare this difference with the volume of the cube you calculated.

Why is the second volume measurement greater than the first? Is there more water?

■ What does this experiment tell you about the fluid an object displaces?

■ Suppose the container of water was filled to the top. What would have happened when you placed the cube into the container?

centimeter (1 g/cm³). The wood block floats because the density of wood is about 0.8 g/cm³. The density of aluminum is 2.7 g/cm³—more than twice the density of water. So the aluminum block sinks. Aluminum can never displace a weight of water equal to its own weight. The density of lead is 11.3 g/cm³. It, too, sinks in water. What happens to lead and aluminum when they are placed in mercury, which has a density of 13.6 g/cm³?

You may wonder how objects such as steel ships are able to float in water, since the density of steel is 7.8 g/cm³. A ship is built of a shell of steel that is hollow inside. So the volume of the ship is made up mostly of air. The ship and air together have a density that is less than that of water. They can displace a weight of water equal to or greater than their weight. Can you explain why a ship will sink if its hull fills with water?

Air is also a fluid. So air exerts a buoyant force. You are buoyed up by the air. But because its buoyant force is so small, you cannot actually feel it. The density of air is only 0.00121 g/cm³. A balloon filled with helium gas will float in air because the density of helium is 0.000167 g/cm³. The density of carbon dioxide is almost twice the density of air. Will a balloon filled with carbon dioxide float in air?

Certain organisms and objects need to float at a certain depth, rather than at the surface. If the weight of a submerged object is exactly equal to the weight of the displaced fluid, the object will not move up or down. Instead, it will float at a constant depth. Most fishes have a gas-filled bladder whose

Figure 3–15 *This bather in the Dead Sea in Israel relaxes with the newspaper as she enjoys the results of density differences. Because the Dead Sea is very salty, it is very dense. Is the woman's density greater or less than that of the water?*

Figure 3–16 *At first glance you may see no relationship between hot-air balloons that rise high in the sky and a submarine that sinks deep into the ocean. But both must adjust their masses to rise, sink, or float at a certain level depending on the buoyant force. Why?*

volume changes to adjust to the buoyant force at various depths. Submerged submarines take on and discharge sea water as needed for the same reason. And the pilot of a hot-air balloon adjusts its weight to match the buoyant force of the air.

The concept of buoyancy is useful in many fields. Geology is a good example. According to the modern theory of plate tectonics and continental drift, the continents can be thought of as floating in a sea of slightly soft rock that acts like a fluid. The height of any continent in a particular area depends in part on the difference between its density and the density of the rock in which it is floating.

3–4 Section Review

1. Why does the buoyant force exist?
2. State Archimedes' principle in terms of buoyancy. In terms of density.
3. The density of ocean water is 1.02 g/cm³. Will a boat float higher in ocean water than in fresh water? Explain your answer.

Critical Thinking—*Designing an Experiment*
4. Suppose a friend finds a strange-looking object with no particular shape. Explain how you can determine the density of the object in order to identify the substance.

ACTIVITY DOING

To Float or Not to Float

1. Obtain a hollow cylinder. A small frozen orange juice can open at both ends will work.

2. Place the cylinder on top of a jar lid.

3. Fill a basin with water. Also fill a drinking glass with water.

4. Hold the jar lid and cylinder together and carefully place them in the basin of water with the lid on the bottom. Hold them so that only about 2 cm of the cylinder are above water.

5. Remove your hand from the jar lid and hold only the cylinder. What happens to the jar lid?

6. Carefully pour water from the drinking glass into the cylinder. What happens to the jar lid when the water level inside the cylinder equals the water level outside?

Explain your observations in steps 5 and 6 in terms of air pressure.

3–5 Fluids in Motion

Now it is time for you to do a little discovering. Try this experiment. Get a long, thin strip of paper. Put the paper in a book by inserting about 5 centimeters of it between two pages. Hold the book upright in front of your mouth so that the paper hangs over the far side of the book as shown in Figure 3–17. Now blow gently across the top of the paper. What happens? If you do this experiment correctly, the piece of paper is pushed upward, or lifted.

What you have demonstrated is the principle formulated by the eighteenth-century Swiss scientist Daniel Bernoulli. **Bernoulli's principle** explains why all forms of flight are possible. **Bernoulli's principle explains that the pressure in a moving stream of fluid is less than the pressure in the surrounding fluid.** The faster a fluid moves, the less pressure it exerts.

When you blow across the top of the paper, you produce a moving stream of air. The pressure in this moving stream is less than the pressure in the surrounding air. So the air pressure under the paper is now greater than the air pressure above it. The paper is pushed up into the moving air. What has this to do with how airplanes fly?

Look at Figure 3–18, which shows the shape of an airplane wing. You will notice that the wing is round in the front, thickest in the middle, and narrow at the back. The bulge in the upper surface makes this surface longer than the lower surface. So when the wing moves forward, the air above the wing must travel a longer distance than the air

Figure 3–17 *You can demonstrate Bernoulli's principle by doing this simple experiment.*

ACTIVITY

DISCOVERING

Rolling Uphill

1. Place a hard-boiled egg or a potato in a small saucepan filled with water.

2. Hold the saucepan under running water so that the water runs between the egg (or potato) and the rim.

3. Tilt the saucepan toward you slightly. Where do you think the egg (or potato) will go? Are you correct?

■ Explain your observations.

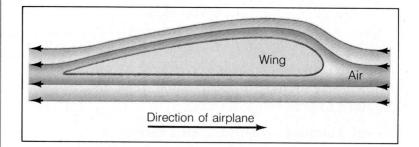

Figure 3–18 *An airplane wing is designed so that air passing over the wing travels faster than air passing beneath it. According to Bernoulli's principle, less pressure is exerted by a fluid that is flowing faster than another fluid. How does this explain how airplanes can fly?*

below the wing. Recalling what you know about speed, what must be true if the air above the wing travels a longer distance in the same amount of time? The air above the wing must be moving faster. According to Bernoulli's principle, then, the air above the wing exerts less pressure on the wing than the air below the wing. This creates an upward unbalanced force that keeps the airplane in the air.

Activity Bank

Baffling With Bernoulli, p.149

PROBLEM Solving

Attack of the Shower Curtain

Smash! You forcefully shut off your alarm clock and drag yourself out of bed. With your eyes only half open, you stumble into the shower. Blindly, you reach over and turn on the water. Aaaah! Suddenly, your eyes pop open as the shower curtain attacks you. You push it away but it comes right back. Instinctively, you shut off the water. To your relief, the shower curtain calms back down. You are a little confused. Could you have imagined all this? You turn the water back on and again the shower curtain moves toward you, climbing on your legs and encircling you. What is going on?

This is no horror story. In fact, it is a common experience you may have had.

Drawing Conclusions

Can you explain why some shower curtains are pulled into the shower with you? What are some ways you can stop the shower curtain from attacking?

Figure 3–19 *Animals that live underground must take advantage of Bernoulli's principle to keep air flowing through their burrows. These prairie dogs, for example, design their mounds so that air pressure above the hole is lower than air pressure inside the burrow. This causes the air inside to be pushed out. What other animals might use similar tactics?*

Figure 3–20 *Up, up, and away goes the beautiful kite as you run vigorously on a windy day. Where is the pressure greater, above or below the kite?*

You can use a simple kite to illustrate the basic ideas of flight. As you run with the kite, the air pushes upward on the kite and the kite rises. And once the kite's weight is balanced by the upward force, the kite remains up. You might want to demonstrate these ideas yourself. Go ahead—fly a kite!

Bernoulli's principle can be used to explain much more than just the flight of an airplane or a kite. You can also use Bernoulli's principle to explain why smoke goes up a chimney. In addition to the fact that hot air rises, smoke goes up a chimney because wind blows across the top of the chimney. This makes the pressure lower at the top of the chimney than at the bottom in the house. This difference in pressure causes the smoke to be pushed up.

3–5 Section Review

1. Explain Bernoulli's principle.
2. How is the shape of an object related to the effects of Bernoulli's principle?
3. Why do airplanes normally take off into the wind?

Critical Thinking—*Relating Cause and Effect*
4. Roofs of houses are sometimes pushed off from the inside by very strong winds. Explain this using Bernoulli's principle.

CONNECTIONS

What a Curve!

It's the bottom of the ninth inning. There are two outs and two strikes on the batter as the pitcher winds up. The batter waits as the ball heads straight for his bat. The batter begins to swing. But wait! In the middle of his swing the ball swerves out of the bat's way. The game is over!

The key to a game-winning *curve ball* is spin. Why do spinning balls curve in flight?

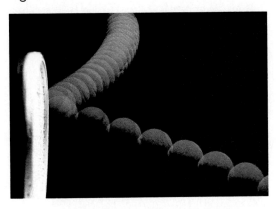

As a spinning ball passes through the air, it drags air around itself in the direction of the spin. For example, consider a tennis ball that is spinning so that the top is being carried in the direction of motion of the ball and the bottom is being carried in the opposite direction. As the ball spins, some of the air will be dragged around in a circular pattern. This will happen as long as the surface of the ball is rough (which is why the fuzz is important). As the ball moves, the air through which it travels appears to move in the opposite direction. Thus the air dragged around the top of the ball is moving in the opposite direction from the air passing the ball.

You know that velocities in opposite directions combine by subtraction. Because of this, the air above the ball slows down. Below the ball, however, the air passing the ball and the air being dragged around the ball are in the same direction. This causes the air below the ball to speed up, as the velocities add together. The air on top of the ball is moving at a slower speed than the air below the ball.

According to Bernoulli's principle, the slower speed means a higher pressure. Thus the air on top of the ball exerts a greater pressure on the ball than the air beneath the ball. This forces the ball downward. A good tennis player can hit the ball with just enough topspin so that it appears to be going out of the court but drops sharply before the baseline.

If a pitcher wishes to make a ball curve, sidespin must be applied to the ball. The direction of the curve will depend on the direction of the sidespin. A spin to the right will curve the ball to the right. A spin to the left will curve the ball to the left. On a baseball, it is the seams that create the airflow around the ball.

So go out and practice Bernoulli's principle for yourself . . . Plaaaay ball!

Laboratory Investigation

A Cartesian Diver

Problem

What is the relationship between the density of an object and its buoyancy in a fluid?

Materials (per group)

copper wire
medicine dropper
large, clear-plastic bottle with
   an airtight lid
glass
water

Procedure 🔬

1. Wrap several turns of wire around the middle of the medicine dropper.

2. Fill the glass with water and place the dropper in the glass. The dropper should barely float, with only the very top of it above the surface of the water.

3. If the dropper floats too high, add more turns of wire. If the dropper sinks, remove some turns of wire.

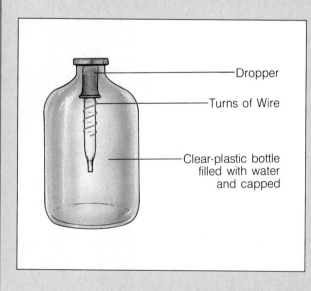

Dropper

Turns of Wire

Clear-plastic bottle
filled with water
and capped

4. Completely fill the large plastic bottle with water.

5. Place the dropper in the bottle of water. The water should overflow.

6. Screw the cap tightly on the bottle. No water or air should leak out when the bottle is squeezed.

7. Squeeze the sides of the bottle. Record your observations. If the dropper does not move, take it out and add more turns of wire.

8. Release the sides of the bottle. Record your observations.

Observations

1. What happens to the dropper when the sides of the bottle are squeezed?

2. What happens to the dropper when the sides of the bottle are released?

Analysis and Conclusions

1. What happens to the pressure of the water when you squeeze the sides of the bottle?

2. When you squeeze the bottle, some of the water is pushed up into the dropper. Why?

3. Why does the dropper sink when you squeeze the sides of the bottle?

4. Why does the dropper rise when you release the sides of the bottle?

5. How is the density of an object related to its buoyancy in a fluid?

6. **On Your Own** Leave the experimental setup in a place where you can observe it at various times of the day. What does it show about air pressure over a period of several days?

Study Guide

Summarizing Key Concepts

3–1 Fluid Pressure

▲ Pressure is a force that acts over a certain area.

▲ The pressure a fluid exerts is due to the fact that the fluid is made up of particles that have mass and motion.

▲ All liquids and gases are fluids. All fluids exert pressure equally in all directions.

▲ Fluids can be pushed from areas of higher pressure to areas of lower pressure.

3–2 Hydraulic Devices

▲ Pressure applied to a fluid is transmitted equally in all directions throughout the fluid.

▲ In hydraulic devices, a small force acting on a small area is multiplied into a larger force acting on a larger area.

3–3 Pressure and Gravity

▲ As a result of gravity, the pressure a liquid exerts increases as the depth increases.

▲ Air pressure decreases as altitude increases.

3–4 Buoyancy

▲ Buoyancy is the phenomenon caused by the upward force of fluid pressure.

▲ The buoyant force on an object is equal to the weight of the fluid displaced by the object. This relationship is called Archimedes' principle.

▲ An object floats in a fluid when the buoyant force on the object is greater than or equal to the weight of the object.

▲ Density is the ratio of the mass of an object to its volume ($D = M/V$).

▲ An object will float in a fluid if its density is less than the density of the fluid.

3–5 Fluids in Motion

▲ Bernoulli's principle states that the pressure in a moving stream of fluid is less than the pressure in the surrounding fluid.

▲ The faster a fluid moves, the less pressure it exerts.

Reviewing Key Terms

Define each term in a complete sentence.

3–1 Fluid Pressure
pressure

3–2 Hydraulic Devices
hydraulic device

3–4 Buoyancy
buoyant force
buoyancy
Archimedes' principle
density

3–5 Fluids in Motion
Bernoulli's principle

Chapter Review

Content Review

Multiple Choice

Choose the letter of the answer that best completes each statement.

1. Force that acts over a certain area is called
   a. density.   c. pressure.
   b. hydraulic.   d. gravity.
2. Pressure in a fluid is exerted
   a. upward only.   c. downward only.
   b. sideways only.   d. in all directions.
3. The weight and motion of fluid particles create
   a. volume.   c. pressure.
   b. mass.   d. density.
4. The pressure of a fluid varies with depth because of
   a. volume.   c. Bernoulli's principle.
   b. gravity.   d. Archimedes' principle.
5. The force of a fluid that pushes an object up is called
   a. hydraulics.   c. buoyancy.
   b. gravity.   d. weight.

6. The relationship between buoyant force and weight of displaced fluid was stated by
   a. Archimedes.   c. Orville Wright.
   b. Newton.   d. Bernoulli.
7. The buoyant force on an object is equal to the weight of the
   a. object.   c. container
   b. displaced fluid. d. entire fluid.
8. Compared with the slow-moving water along the edge of a river, the rapidly-moving stream in the middle exerts
   a. less pressure.   c. more pressure.
   b. no pressure.   d. the same pressure.
9. When compared with the air that travels under an airplane wing, the air that travels over the wing
   a. is more dense.   c. moves more slowly.
   b. is less dense.   d. moves faster.

True or False

If the statement is true, write "true." If it is false, change the underlined word or words to make the statement true.

1. Pressure is force per unit <u>mass</u>.
2. Fluids will move from areas of <u>high</u> pressure to areas of <u>low</u> pressure.
3. Pressure varies with depth due to the force of <u>gravity</u>.
4. The force of a fluid that pushes an object up is called <u>buoyant</u> force.
5. The buoyant force on an object equals the <u>volume</u> of the displaced fluid.
6. An object will float in a fluid whose density is <u>less</u> than the density of the object.
7. Pressure in a moving stream of fluid is <u>greater than</u> the pressure in the surrounding fluid.
8. The flight of an airplane can be explained using <u>Bernoulli's</u> principle.

Concept Mapping

Complete the following concept map for Section 3–1. Refer to pages S6–S7 to construct a concept map for the entire chapter.

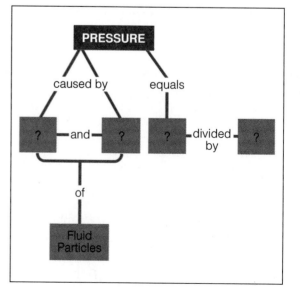

Concept Mastery

Discuss each of the following in a brief paragraph.

1. What is fluid pressure?
2. Explain why an astronaut must wear a pressurized suit in space.
3. Heating a fluid causes its pressure to increase. Why should you poke holes in the skin of a potato before it is baked?
4. Using the principle of fluid pressure, explain how a medicine dropper works.
5. Describe how a hydraulic device works.
6. Why does fluid pressure increase as depth increases?
7. What is the effect of increased water depth on a scuba diver?
8. Explain why you weigh more in air than you do in water.
9. Why does the canvas top of a convertible car bulge out when the car is traveling at high speed?
10. Hummingbirds expend 20 times as much energy to hover in front of a flower as they do in normal flight. Explain.

Critical Thinking and Problem Solving

Use the skills you have developed in this chapter to answer each of the following.

1. **Applying concepts** Air exerts a downward force of 100,000 N on a tabletop, producing a pressure of 1000 N/cm^2.
   a. What would be the force if the tabletop were twice as large?
   b. What would be the pressure if the tabletop were twice as large?
2. **Designing an experiment** The density of gold is 19.3 g/cm^3. The density of pyrite, or fool's gold, is 5.02 g/cm^3. Using mercury, density 13.6 g/cm^3, describe an experiment by which you could tell the difference between samples of the two substances.
3. **Applying concepts** Explain why salad dressing made of oil and vinegar must be shaken before use.
4. **Applying concepts** Describe how you could make a sheet of aluminum foil float in water. How could you change its shape to make it sink?
5. **Applying concepts** A barge filled with sand approaches a bridge over the river and cannot quite pass under it. Should sand be added to or removed from the barge?
6. **Applying concepts** A student holds two sheets of paper a few centimeters apart and lets them hang down parallel to each other. Then the student blows between the two papers. What happens to the papers? Why?
7. **Relating facts** A Ping-Pong ball can be suspended in the air by blowing a stream of air just above it. Explain how this works.
8. **Identifying relationships** Airplanes are riveted together at the seams. The rivets are installed so they are even with the outside surface. Why is it important that the outside surface be so smooth?
9. **Using the writing process** It's your first assignment as a cub reporter. You are to interview the Wright brothers after their pioneering flight. Make a list of five questions you would ask.

Work, Power, and Simple Machines

Guide for Reading

After you read the following sections, you will be able to

4–1 What It Means to Do Work
- Relate force, work, and distance.
- Calculate work.

4–2 Power
- Calculate power.

4–3 Machines
- Describe the role of machines in doing work.

4–4 Simple and Compound Machines
- Name the six simple machines.
- Show how simple machines are related to compound machines.

The Great Pyramid of Khufu in Egypt is one of the Seven Wonders of the World. It stands over 137 meters high. Its base covers an area large enough to hold ten football fields. More than 2 million stone blocks, each weighing about 20,000 newtons (about the weight of two large cars), make up its structure.

The Great Pyramid is a tribute to human effort and ingenuity. For it is exactly these two qualities that enabled the Egyptians to chisel the stone blocks from limestone quarries, to transport them to the pyramid site, and to raise them to the top of the magnificent structure.

The Egyptians had only simple machines with which to work. Their only source of power was human effort. Several hundred thousand people toiled for twenty years to build the Great Pyramid. Today, with modern machinery, it could be built with only a few hundred workers and in one fifth the time!

In this chapter you will learn about work, power, and simple machines. And you will gain an understanding of how machines make work easier—certainly easier than it was for the Egyptians who built the Great Pyramid.

Journal *Activity*

You and Your World Doorknobs, wheels, shovels, ramps, screwdrivers, and buttons are but a few examples of the many tools and machines you use each day. In your journal, explore how different your day would be if you lived during the time in which the pyramids were built—a time in which most of these devices did not exist. What devices would you miss most?

The Great Pyramid of Khufu stands at Giza, Egypt.

4–1 What It Means to Do Work

People use the word work in many different ways. You say that you work when you study for a test. A lifeguard may use the word work to describe watching people swim in a pool or lake. A television newscaster thinks of work as reporting the news. All these people believe they are doing work. But a scientist would not agree!

The term **work** has a special meaning in science. Work is done only when a force moves an object. When you push, lift, or throw an object, you are doing work. Your force is acting on the object, causing it to move a certain distance. **A force acting through a distance is work.** You do work whenever you move something from one place to another.

Work is not done every time a force is applied, however. A force can be exerted on an object without work being done on the object. How can this be so? Suppose you push as hard as you can against a wall for several minutes. Obviously, the wall does not move. And although you may be extremely tired, you have not done any work. According to the definition of work, a force must be exerted over a distance. Although you have applied a force to the wall, the wall did not move. Thus no work has been done.

Figure 4–1 *A scientist knows that work is done only when a force moves an object through a distance. The karate student is doing work; the lifeguards are not. Why?*

Figure 4–2 *In the scientific sense, why is no work being done by the person's arms in carrying the bag of groceries? Why is work being done in lifting the bag?*

Figure 4–3 *Everyone knows it's hard work being a mother—in more ways than one. Just observe this mother chimpanzee as she carries her baby, who has decided to take a free ride.*

Another important requirement for work to be done is that the distance the object moves must be in the same direction as the force applied to the object. Let's see just what this means. Imagine that you are given a heavy bag of groceries to carry. Your muscles exert an upward force on the bag in order to hold it up. Now suppose you walk toward the door. Have your arms done any work on the bag of groceries? The answer is no. The direction of movement of the bag is not the same as the direction of the applied force. The applied force is upward, whereas the direction of movement is forward. What would you have to do with the bag in order to do work on it?

The amount of work done in moving an object is equal to the force applied to the object times the distance through which the force is exerted (the distance the object moves):

Work = Force × Distance

Force is measured in newtons. Distance is measured in meters. So the unit of work is the newton-meter (N-m). In the metric system, the newton-meter is called the **joule** (J). A force of 1 newton exerted on an object that moves a distance of 1 meter does 1 newton-meter, or 1 joule, of work.

If you lifted an object weighing 200 N through a distance of 0.5 m, how much work would you do? The force needed to lift the object up must be equal to the force pulling down on the object. This is the object's weight. So the force is 200 N. The amount of work is equal to 200 N × 0.5 m, or 100 J.

ACTIVITY

CALCULATING

It Takes Work to Catch a Flight

There was an announcement made over the loudspeaker. The flight to Los Angeles was about to depart. A 600-newton woman who was waiting for the flight lifted her 100-newton suitcase a distance of 0.5 meter above the airport floor and ran 25 meters.

Calculate how much work was done by the woman's arms in moving the suitcase. Draw a diagram showing the forces and distances involved in this situation.

Explain how the work done would change if she had dragged her suitcase along horizontally instead of lifting it. Draw a diagram showing this situation.

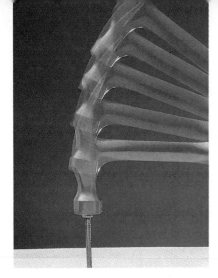

Figure 4–4 *A hammer pounding a nail exerts a force on the nail causing it to move a certain distance into the wood. Is work done on the hammer?*

4–1 Section Review

1. What is work?
2. How are force, work, and distance related?
3. What are the units of work?
4. A 900-N mountain climber scales a 100-m cliff. How much work is done by the mountain climber?

Critical Thinking—*Analyzing Information*

5. A small turtle slowly crawls along carrying a fallen bird feather on its back. After quite some time, the turtle passes an elephant standing still with five large lions on its back. Who is doing more work, the turtle or the elephant? Explain.

Guide for Reading

Focus on this question as you read.

▶ *How is power related to work and time?*

4–2 Power

The word **power** is like the word "work." It has different meanings to different people. But in science, power has a very specific meaning. Like speed, velocity, and acceleration, power tells you how fast something is happening—in this case, how fast work is being done. **Power is the rate at which work is done, or the amount of work per unit of time.**

Power, then, is calculated by dividing the work done by the time it takes to do it:

$$\text{Power} = \frac{\text{Work}}{\text{Time}}$$

Since work equals force times distance, the equation for power can also be written:

$$\text{Power} = \frac{\text{Force} \times \text{Distance}}{\text{Time}}$$

The unit of power is simply the unit of work divided by a unit of time, or the joule per second. This unit is also called a **watt** (W). One watt is equal to 1 joule per second (1 J/sec).

You are probably familiar with the watt as it is used to express electric power. Electrical appliances and light bulbs are rated in watts. A 50-watt light

Figure 4–5 *Both the man and the snowplow are doing work, but there is little doubt that the machine is doing more work in the same amount of time. So the machine has more power. How is power calculated?*

bulb does work at the rate of 50 joules per second. In the same second, a 110-watt light bulb does 110 joules of work. The 110-watt light bulb is more powerful than the 50-watt light bulb. Large quantities of power are measured in kilowatts (kW). One kilowatt equals 1000 watts. The electric company measures the electric power you use in your home in kilowatts.

Perhaps you can now see why a bulldozer has more power than a person with a shovel. The bulldozer does more work in the same amount of time. As the process of doing work is made faster, power is increased. Can you explain why it takes more power to run up a flight of stairs than it takes to walk up?

4–2 Section Review

1. What is power? What is the relationship among power, work, and time?
2. What is a watt?
3. A small motor does 4000 J of work in 20 sec. What is the power of the motor in watts?

Critical Thinking—*Relating Concepts*

4. Suppose you ride in a sleigh being pulled by horses at 16 kilometers per hour. Another sleigh being pulled at 10 kilometers per hour travels the same distance you do. Which horses are more powerful? How is speed related to power?

ACTIVITY
DOING

Work and Power

1. Determine your weight in newtons. (Multiply your weight in pounds by 4.5).

2. Determine how many seconds it takes you to walk up a flight of stairs.

3. Determine how many seconds it takes you to run up the same flight of stairs. Be careful as you run.

4. Measure the vertical height of the stairs to the nearest 0.01 meter.

5. Using the formula work = weight x height, calculate the work done in walking and running up the stairs.

6. Calculate the power needed for walking and for running up the stairs.

Is there a difference in the work done in walking and running? In the power?

The Power of Nature

■ Nebraska: Buildings are blown apart. Houses and cars are thrown about like toys. A beautiful town is leveled in a matter of minutes.

■ New York: A defenseless city is crippled by a two-foot blanket of snow.

■ Florida: Tall, stately palm trees are bent over as far as they can go. Fierce winds break windows and knock off roofs while sheets of rain batter coastal cities.

■ Ohio: A rural area is evacuated as flood waters approach.

What force is powerful enough to cause so much destruction? What force is capable of doing the work required to move people, buildings, and water? As you may have already guessed, the source of this awesome power is nature!

Powerful *weather* conditions are a result of the interactions of several factors in the Earth's atmosphere. In particular, changes in weather are caused by movements of air masses. An air mass is a large volume of air that has the same temperature and contains the same amount of moisture throughout. When two different air masses meet, an area called a front is created. The weather at a front is usually unsettled and stormy. And when two fronts collide—watch out! The results are rainstorms, thunderstorms, hail storms, or snowstorms, depending on the characteristics of the various air masses.

Weather affects people daily and it influences them and the world around them. The type of homes people build, the clothes they wear, the crops they grow, and the jobs they work at are all determined by the weather. It is important to appreciate the fact that despite all of our technological advances, the greatest source of power is still found in nature. And when this power is destructive, or even inconvenient, we are ultimately defenseless.

4–3 Machines

Guide for Reading

Focus on this question as you read.

▶ *How do machines affect work?*

Think about performing some common activities without the devices you normally use. Consider eating soup without a spoon, opening the door without a doorknob, removing snow without a shovel, cutting the lawn without a lawn mower. For centuries, people have looked for ways to make life more enjoyable by using devices that make work easier.

An instrument that makes work easier is called a **machine.** Machines are not limited to the complicated devices you may be thinking of—car engines, airplanes, computers, factory equipment. In fact, some machines do not even have moving parts. A machine is any device that helps you to do something.

How Do Machines Make Work Easier?

There are always two types of work involved in using a machine: the work that goes into the machine and the work that comes out of the machine. The work that goes into the machine is called the **work input.** The work input comes from the force that is applied to the machine, or the effort force. When you use a machine, you supply the

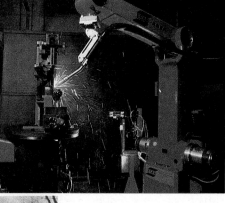

Figure 4–6
Machines have come a long way—from simple water wheels to automated robots. Regardless of their complexity, all machines have the same purpose: to make work easier.

ACTIVITY READING

Science in a Chocolate Factory

You may have seen or heard of a movie about a zany inventor called Willie Wonka and his fantastic chocolate factory. In this factory, Wonka creates a collection of wonderful machines that perform all sorts of tasks—including making bubble gum. The movie was based on an equally entertaining book called *Charlie and the Chocolate Factory* by Roald Dahl. You might enjoy reading this delightful story.

effort force. Because you exert this force over some distance, you put work into the machine.

Of course the machine does work too. The machine exerts a force, called an output force, over some distance. The work done by the machine is called the **work output.** The work output is used to overcome the force you and the machine are working against. This force, which opposes the effort force, is called the resistance force. The resistance force is often the weight of the object being moved. For example, when you use a shovel to move a rock, your effort is opposed by the rock's weight. The rock's weight is the resistance force.

Machines do not increase the work you put into them. This is a very important idea that you should keep in mind as you continue to read about machines. The work that comes out of a machine can never be greater than the work that goes into the machine. Like momentum, work is conserved. Why, then, do we say that machines make work easier?

What machines do is change the factors that determine work. **Machines make work easier because they change either the size or the direction of the force put into the machine.** Any change in the size of a force is accompanied by a change in the distance through which the force is exerted. If a machine multiplies the force you put into it, the output force (the force you get out of a machine) will be exerted over a shorter distance. If a machine exerts an output force over a longer distance than the

Figure 4–7 *A machine can make a task easier in one of three ways. It can multiply the size of the force, but decrease the distance over which the force moves. It can multiply the distance over which the force moves, but decrease the size of the force. Or it can leave both force and distance unchanged, but change the direction in which the force moves.*

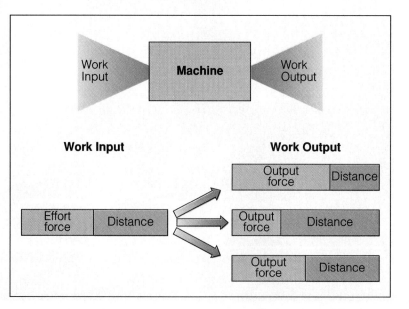

Figure 4–8 *Believe it or not, a merry-go-round is a type of machine. The children who push the merry-go-round provide the effort force. The output force spins the children riding on the merry-go-round in circles.*

effort force, the output force will be less than the effort force. Because work is conserved (it does not change), what you increase in force you pay for in distance, and what you increase in distance is at the expense of force. In other words, most machines make work easier by multiplying either force or distance—but never both. No machine can multiply both force and distance!

Determining How Helpful a Machine Is

As you just learned, the work that comes out of a machine (work output) can never be greater than the work that goes into the machine (work input). In reality, the work output is always less than the work input. Do you know why? The operation of any machine always involves some friction. Some of the work the machine does is used to overcome the force of friction. Scientists have a way of comparing the work output of a machine to its work input— and thereby knowing how much work is lost to friction. The comparison of work output to work input is called the **efficiency** of a machine. The closer work output is to work input, the more efficient the machine. What does this mean in terms of friction?

Efficiency is expressed as a percentage. Efficiency can never be greater than 100 percent, because the work output can never be greater than the work input. In fact, there is no machine that has an efficiency of 100 percent. Machines with the smallest

Figure 4–9 *Many attempts have been made to create a machine with 100 percent or more efficiency—a perpetual motion machine. So far, this has proved impossible. What force reduces the efficiency of a machine?*

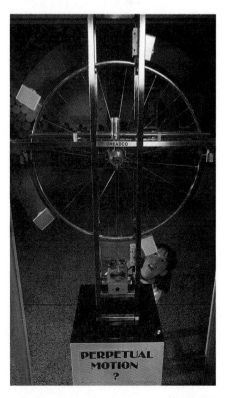

Figure 4–10 *Many household appliances come with guides that inform potential customers of the efficiency of the appliance. A more efficient appliance will be easier to run and will save the owner money on the electric bill.*

amount of friction are the most efficient. For this reason, it is important to keep a machine well lubricated and in good condition.

In addition to knowing how efficient a machine is, we can also determine how helpful a machine is. What we mean by helpful is how many times the machine multiplies the effort force to overcome the resistance force. The number of times a machine multiplies the effort force is called the **mechanical advantage** of the machine. The mechanical advantage tells you how much force is gained by using the machine. The more times a machine multiplies the effort force, the easier it is to do the job.

4–3 Section Review

1. What is a machine?
2. What advantage is there to using a machine if it does not multiply the work put into it?
3. What is effort force? Resistance force?
4. What is efficiency? Mechanical advantage?

Critical Thinking—*Drawing Conclusions*
5. Describe the relationship between friction and the efficiency of a machine.

4–4 Simple and Compound Machines

The devices you think of when you hear the word "machine" are actually combinations of two or more simple machines. **There are six types of simple machines: the inclined plane, the wedge, the screw, the lever, the pulley, and the wheel and axle.**

Inclined Plane

Suppose you had to raise a car to a height of 10 centimeters. How would you do it? Certainly not by lifting the car straight up. But you could do it if you pushed the car up a ramp. A ramp would make the job easier to do. Although a ramp does not alter the amount of work that is needed, it does alter the way in which the work is done. A ramp decreases the amount of force you need to exert, but it increases the distance over which you must exert your force. Remember, what you gain in force, you pay for in distance.

A ramp is an example of an **inclined plane.** An inclined plane is simply a flat slanted surface. An inclined plane is a simple machine with no moving parts. The less slanted the inclined plane, the longer the distance over which the effort force is exerted and the more the effort force is multiplied. Thus the mechanical advantage of an inclined plane increases as the slant of the plane decreases. The principle of the inclined plane was of great importance in ancient times. Ramps enabled the Egyptians to build their pyramids and temples. Since then, the inclined plane has been put to use in many devices from door locks to farming plows to zippers.

WEDGE In many devices that make use of the inclined plane, the inclined plane appears in the form of a **wedge.** A wedge is an inclined plane that moves. In a wedge, instead of an object moving along the inclined plane, the inclined plane itself moves to raise the object. As the wedge moves a greater distance, it raises the object with greater force. A wedge is usually a piece of wood or metal that is thinner at one end. Most wedges are made up of two inclined planes. A knife and an ax are two examples.

The longer and thinner a wedge is, the less the effort force required to overcome the resistance force. (This is true of the inclined plane as well.) When you sharpen a wedge, you are increasing its mechanical advantage by decreasing the effort force that must be applied in using it. A sharpened ax requires less effort force because the edge is thinner.

A lock is another device that depends on the principle of the wedge. Think for a moment about the shape of a key. The edges go up and down in a certain pattern. The edges of a key are a series of wedges. The wedges lift up a number of pins of different lengths inside the lock. When all of the pins are lifted to the proper height, which is accomplished by the shape of the key, the lock opens.

Figure 4–11 *An inclined plane is a slanted surface used to raise an object. An inclined plane decreases the size of the effort force needed to move an object. What happens to the distance through which the effort force is applied?*

Figure 4–12 *As a wedge is moved through an object to be cut, a small effort force is able to overcome a large resistance force. How can the mechanical advantage of a wedge be increased?*

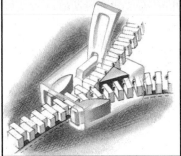

Figure 4–13 *A plow is a device that uses a combination of wedges to cut, lift, and turn over the soil. You have probably never given much thought to the zipper that fastens your clothes. But zippers consist of wedges. One wedge is used to open the zipper. Two different wedges are used to push the two sides together.*

The zipper is another important application of the wedge. Zippers join or separate two rows of interlocking teeth. Have you ever tried to interlock the two sides of a zipper with your hands? It is almost impossible to create enough force with your fingers to join the two rows of teeth. However, the part of the zipper that you pull up or down contains three small wedges. These wedges turn the weak effort force with which you pull into a strong force that either joins or separates the two sides. Without these wedges, you would not be able to use the zipper.

SCREW Just as the wedge is an inclined plane that moves, the **screw** is an inclined plane wrapped around a central bar, or cylinder, to form a spiral. A screw rotates, and with each turn moves a certain distance up or down. A screw multiplies an effort force by acting through a long distance. The closer

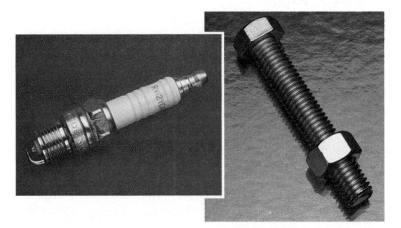

Figure 4–14 *Screws come in a variety of shapes and sizes. A bolt and a spark plug are but two examples. How is a screw related to an inclined plane?*

together the threads, or ridges, of a screw, the longer the distance over which the effort force is exerted and the more the force is multiplied. Thus the mechanical advantage of a screw increases when the threads are closer together.

The wood screw and the corkscrew are two obvious examples of the screw. Another example is a nut and bolt. In a nut and bolt, the nut has to turn several times to move forward a short distance. However, the nut moves along the bolt with a much greater force than the effort force used to turn it. Faucets and jar lids also take advantage of the principle of a screw. It is important for a jar lid to close tightly, but it requires a great deal of force to achieve a tight seal. Inside a jar lid are the threads of a screw. They fit into those on the top of the jar. You exert a small effort force when you turn the lid. However, your effort force is multiplied because you exert it over a long distance as you turn the lid many times. The large output force seals the jar.

Lever

Have you ever ridden on a seesaw or pried open a can with a screwdriver? If so, you are already familiar with the simple machine called a **lever.** A lever is a rigid bar that is free to pivot, or move about, a fixed point. The fixed point is called the **fulcrum.** When a force is applied on a part of the bar by pushing or pulling it, the lever swings about the fulcrum and overcomes a resistance force. Here is an example. Let's suppose you are using a crowbar to remove a nail from a piece of wood. When you push down on one end of the crowbar, the nail moves in the other direction—in this case, up. The crowbar changes the direction of the force. But the force exerted on the nail by the crowbar moves a shorter distance than the effort force you exert on the crowbar. In other words, you push down through a longer distance than the nail moves up. Because work is conserved, this must mean that the crowbar multiplies the effort force you apply.

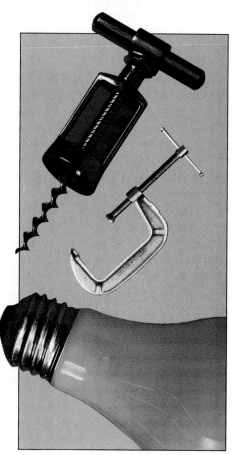

Figure 4–15 *These screws increase the amount of force applied in turning them, but decrease the distance over which the force is applied.*

Figure 4–16 *An essential attraction of any playground, a seesaw, is a lever. Why is a seesaw a simple machine?*

First-Class Lever	Second-Class Lever	Third-Class Lever

Figure 4–17 *The relative positions of the effort force, resistance force, and fulcrum determine the three classes of levers. Which lever multiplies effort force as well as changes its direction? Which lever multiplies the distance of the effort force?*

Figure 4–18 *One person may be able to lift a large rock with a sturdy lever. The same rock, however, would require a much greater effort force if the length of the lever was shortened. Should the effort force be applied closer to or farther away from the fulcrum to make a chore easier?*

In the case of the crowbar, the fulcrum is between the effort force (your push) and the resistance force (the nail). The fulcrum of a lever is not always between the effort force and the resistance force, however. Sometimes it is at the end of the lever. In fact, levers are divided into three groups, or classes, depending on the location of the fulcrum and the forces. Levers such as a crowbar, seesaw, and pliers are in the first class.

A wheelbarrow is a second-class lever. In a wheelbarrow, the fulcrum is at the end. The wheel acts as the fulcrum. The resistance force is the weight of the load in the wheelbarrow (often rock or soil). And the effort force, which is at the other end of the lever, is the force that you apply to the handles to lift the wheelbarrow. In this case, the effort force is exerted over the distance you lift the handles. The load moves a much shorter distance than you actually lift the handles. Keeping in mind that work is always conserved, you can see that because distance is decreased by the wheelbarrow, force must be increased. So you must exert a smaller force with

Figure 4–19 *"Row. Row. Row" shouts the captain as the crew team sweats through an early morning practice. Although they may hardly be aware of it, their work is made much easier by the action of levers. Where are the levers in this photo?*

the wheelbarrow than you would need to if you lifted the soil directly, but you must exert your force for a longer distance. Like most first-class levers, levers in this second class multiply force but decrease distance. In this second class of levers, however, the direction in which you lift is the same as the direction in which the load moves. A second-class lever does not change the direction of the force applied to it. Doors, nutcrackers, and bottle openers are additional examples of second-class levers. Can you explain why by identifying the location of the fulcrum and forces?

A fishing pole is a third-class lever. The fixed point, or fulcrum, is at the end of the pole where you are holding it. The effort force is applied by your other hand as you pull back on the pole. At the top of the pole is the resistance force. In this case, you need to move your effort force only a short distance to make the end of the pole move a greater distance. A lever in the third class reduces the effort force required but multiplies the distance through which the output force moves. Shovels, hoes, hammers, tweezers, and baseball bats are third-class levers.

Have you ever tried to pry open a can of paint with a short stick? If so, you know that it is rather difficult to do. But what if you use a longer stick or even a screwdriver? The task becomes much easier. What this example illustrates is that where you push or pull on a lever is just as important as the amount of force you apply to it. Less effort force can move the same load if that force is applied farther

ACTIVITY

DOING

Levers

1. Tape five large washers together to form a resistance force.

2. Obtain a rigid 30-centimeter (cm) ruler to use as a lever and a pen to use as a fulcrum.

3. Place the washers on the 1-cm mark. Place the pen under the ruler at the 10-cm mark.

4. Push down on the ruler at the 30-cm mark. Your push is the effort force.

5. Move the pen to the 15-cm mark and again push down at the 30-cm mark.

6. Compare your effort force in steps 4 and 5.

7. Move the pen to the 20-cm mark and again push down on the 30-cm mark.

What is the effect on the effort force of decreasing the distance between it and the fulcrum? What class lever is this?

away from the fulcrum. As long as the resistance force (the load) is closer to the fulcrum than the effort force, the lever will multiply the effort force. This is true for first- and second-class levers. For third-class levers, however, the resistance force is always farther away from the fulcrum than the effort force. So these levers cannot multiply force. But this should not surprise you because you learned that a fishing pole does not multiply force, only distance.

Some familiar instruments are examples of combinations of levers. A pair of scissors, for example, is a combination of two first-class levers. The fulcrum is the center of the scissors, where the two blades are connected. The object to be cut exerts the resistance force. The effort force is exerted by the person using the scissors.

A grand piano is another example of levers working together. Each key on the piano is linked to a complex system of levers. The levers transmit movement from the player's fingers to the felt-tipped hammer, which strikes the tight piano wire and sounds a note. These levers multiply movement so that the hammer moves a greater distance than the player's fingertips. Similarly, a manual typewriter converts a small movement of the fingertip into a long movement of the key to the paper.

Pulley

Have you ever raised or lowered a window shade? If so, you were using another type of simple machine, the **pulley.** A pulley is a rope, belt, or chain wrapped around a grooved wheel. A pulley can function in two ways. It can change the direction of a force or the amount of force.

Suppose you had to lift a very heavy object by yourself. How would you do it? It would be quite difficult, if not impossible, to lift the object directly off the ground. It would be easier to attach the object to a rope that moves through a pulley attached to the ceiling, and then pull down on the rope. After all, pulling down on a rope is a lot easier than lifting the object directly up. By using a pulley, you can change the direction in which you have to apply the force. A pulley that is attached to a structure is called a fixed pulley. A fixed pulley does not multiply

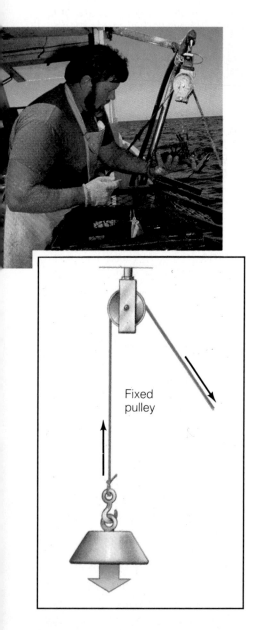

Figure 4–20 *A fixed pulley makes lifting an object easier by changing the direction of the effort force. These unlucky lobsters have been hoisted up with the use of a fixed pulley.*

Fixed pulley

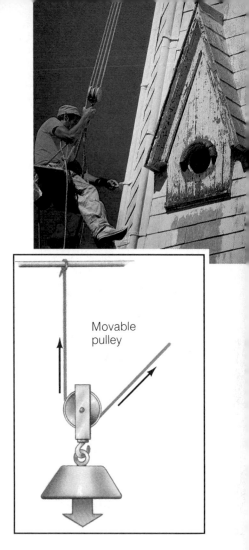

Figure 4–21 *A movable pulley moves with the resistance force. Do movable pulleys change the direction of the effort force?*

Movable pulley

an effort force. It only changes the direction of the effort force. The fact that the effort force is not changed tells you something about the distance the object moves. Because work output can never be greater than work input, if the pulley does not change the amount of force, it does not change the distance the force moves. So the distance you pull is the same as the distance the object moves.

Pulleys can be made to multiply the force with which you pull on them. This is done by attaching a pulley to the object you are moving. This type of pulley is called a movable pulley. Look at the pulley in Figure 4–21. For each meter the load moves, the force must pull two meters. This is because as the load moves, both the left and the right ropes move. Two ropes each moving one meter equals two meters. So although a movable pulley multiplies the effort force, you must exert that effort force for a greater distance than the distance the output force moves the object. Movable pulleys, however, cannot change the direction of an effort force.

A greater mechanical advantage can be obtained by combining fixed and movable pulleys into a pulley system. As more pulleys are used, more sections of rope are attached to the system. Each additional section of rope helps to support the object. Thus less force is required. This increases the mechanical advantage. A block and tackle is a pulley system.

PULLEY SYSTEMS

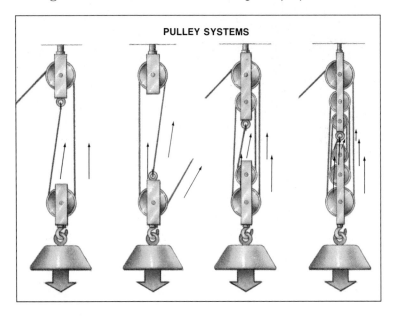

Figure 4–22 *By combining fixed and movable pulleys into a pulley system, the mechanical advantage is increased. The block and tackle on this boat is an example of a pulley system.*

ACTIVITY

*Simple Machines
Around You*

1. Make a data table with six columns. Head each column with one of the six simple machines.

2. Walk around your house, garage, yard, and school. Identify all of the simple machines you see. Record your observations on the data table.

3. Extend your observations by noticing simple machines in other locations, such as a department store, supermarket, bank, and playground.

Which is the most common simple machine? The least common?

Wheel and Axle

Do you think you would be able to insert a screw into a piece of wood using your fingers? Why not try it and see. You will find it is almost impossible to do—which is precisely why you use a screwdriver! A screwdriver enables you to turn the screw with relative ease. A screwdriver is an example of a tool that uses the principle of a **wheel and axle.**

A wheel and axle is a simple machine made up of two circular objects of different sizes. The wheel is the larger object. It turns about a smaller object called the axle. Because the wheel is larger than the axle, it always moves through a greater distance than the axle. A force applied to the wheel is multiplied when it is transferred to the axle, which travels a shorter distance than the wheel. Remember, work must remain the same. The mechanical advantage depends on the radius of the wheel and of the axle. If the radius of the wheel is four times greater than the radius of the axle, every time you turn the wheel once, your force will be multiplied by four.

Bicycles, Ferris wheels, gears, wrenches, doorknobs, and steering wheels are all examples of wheels and axles. Water wheels also use the principle of a wheel and axle. The force of the water on the

Figure 4-23 *A youngster enjoys a ride with the help of several wheels and axles. Can you identify them? Applying a small effort force to the wheel causes it to move through a greater distance than the axle. Thus the force is multiplied at the axle. The same principle is responsible for the operation of a wheelchair and of gears.*

paddles at the rim produces a strong driving force at the central shaft. Windmills use sails to develop power from the wind. The force of wind along the sails produces a strong driving force at the central shaft. A less pleasant but direct descendant of the windmill is the dentist's drill, which uses a stream of air to turn the drill.

Compound Machines

A car is not one of the six simple machines you have just learned about. Rather, a car is a combination of simple machines: wheels and axles; a gearshift lever; a set of transmission gears; a brake lever; and a steering wheel, to name just a few.

Cars, bicycles, watches, and typewriters are all examples of compound machines. Most of the machines you use every day are compound machines. **A compound machine is a combination of two or more simple machines.**

You are surrounded by a great variety of compound machines. How many compound machines do you have in your home? A partial list might include a washing machine, VCR, blender, sewing machine, and vacuum cleaner. Compound machines make doing work easier and more enjoyable. But remember that machines, simple or compound, cannot multiply work. You can get no more work out of a machine than you put into it!

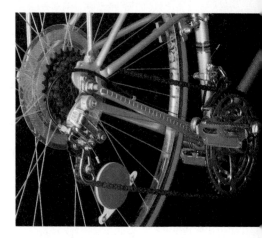

Figure 4–24 *The wheel and gears of a bicycle consist of a combination of simple machines. How many can you identify?*

Compound Machines

Design a machine that uses all six simple machines. Your machine should do something useful, like wash a pet snail or scratch your back. Draw a diagram or build a model of your machine. Label each simple machine. Accompany your model with a short written explanation of what your machine does.

4–4 Section Review

1. Describe the six simple machines.
2. How is slant related to the mechanical advantage of an inclined plane, wedge, and screw?
3. How can you increase the mechanical advantage of a wheel and axle? A lever?

Connection—*You and Your World*
4. An elevator is continously lifted up and lowered down. Which of the six simple machines would be most important to the operation of an elevator? Explain your reasoning.

Laboratory Investigation

Up, Up, and Away!

Problem

How do pulleys help you to do work?

Materials *(per group)*

> ring stand and ring
> spring balance calibrated in newtons
> weight
> string
> single pulley

Procedure

1. Tie one end of a small piece of string around the weight. Tie the other end to the spring balance. Weigh the weight. Record the weight in newtons. Untie the string and weight.
2. Attach the ring about one half to three fourths of the way up the ring stand.
3. To construct a single fixed pulley, hang the pulley directly onto the ring as shown.
4. Tie the weight to one end of a string.
5. Pass the other end of the string over the pulley and tie it to the spring balance.
6. Pull down slowly and steadily on the spring balance and record the force needed to raise the weight.

7. To make a single movable pulley, tie one end of a string to the ring.
8. Pass the other end of the string under the pulley and tie it to the spring balance as shown.
9. Attach the weight directly onto the pulley.
10. Raise the weight by pulling the spring balance upward. Record the force shown on the spring balance.

Observations

1. How much force was needed to lift the weight using the fixed pulley?
2. How much force was needed to lift the weight using the movable pulley?

Analysis and Conclusions

1. How does a fixed pulley help you do work?
2. How does a movable pulley help you do work?
3. What could you do to lift an object with greater ease than either the fixed pulley or the movable pulley alone?
4. **On Your Own** Using what you learned about pulleys, figure out how many movable pulleys you would need to lift a 3600-N boat using a force of 450 N.

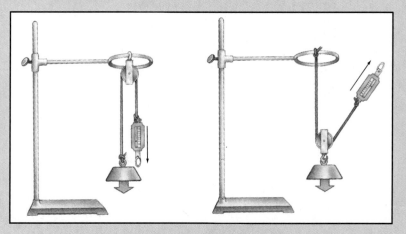

Summarizing Key Concepts

4–1 What It Means to Do Work

▲ Work is the product of force applied to an object times the distance through which the force is applied.

▲ The metric unit of work is the joule.

4–2 Power

▲ Power is the rate at which work is done.

▲ The metric unit of power is the watt.

4–3 Machines

▲ A machine changes either the size or direction of an applied force.

▲ Effort force is force applied to a machine. Work put into a machine is work input.

▲ Work that comes out of a machine is work output. Work output overcomes the resistance force.

▲ The comparison of work output to work input is called efficiency.

▲ The mechanical advantage of a machine is the number of times the machine multiplies the effort force.

4–4 Simple and Compound Machines

▲ There are six simple machines: the inclined plane, the wedge, the screw, the lever, the pulley, and the wheel and axle.

▲ The inclined plane is a slanted surface.

▲ The wedge is a moving inclined plane.

▲ The screw is an inclined plane wrapped around a cylinder.

▲ The lever is a rigid bar that is free to move about the fulcrum when an effort force is applied. There are three classes of levers depending upon the locations of the fulcrum, the effort force, and the resistance force.

▲ A pulley is a chain, belt, or rope wrapped around a grooved wheel. A fixed pulley changes the direction of an effort force. A movable pulley multiplies the effort force.

▲ A wheel and axle is a simple machine made up of two circular objects with different diameters.

▲ A compound machine is a combination of two or more simple machines.

Reviewing Key Terms

Define each term in a complete sentence.

4–1 What It Means to Do Work
work
joule

4–2 Power
power
watt

4–3 Machines
machine
work input
work output

efficiency
mechanical advantage

4–4 Simple and Compound Machines
inclined plane
wedge
screw
lever
fulcrum
pulley
wheel and axle

Chapter Review

Content Review

Multiple Choice

Choose the letter of the answer that best completes each statement.

1. Even if a large force is exerted on an object, no work is performed if
   a. the object moves.
   b. the object does not move.
   c. the power is too large.
   d. the power is too small.
2. The rate at which work is done is called
   a. energy.   c. power.
   b. efficiency.   d. mechanical advantage.
3. Two forces always involved in using a machine are
   a. effort and fulcrum.
   b. friction and fulcrum.
   c. resistance and wattage.
   d. effort and resistance.
4. The comparison between work output and work input is
   a. power.   c. mechanical advantage.
   b. efficiency.   d. friction.
5. Decreasing the slant of an inclined plane increases its
   a. mechanical advantage.
   b. effort force.
   c. power.
   d. work output.
6. The effort force is multiplied in a
   a. corkscrew.   c. baseball bat.
   b. fixed pulley.   d. fishing pole.
7. An example of a second-class lever is a
   a. seesaw.   c. door.
   b. shovel.   d. crowbar.
8. Neither force nor distance is multiplied by a (an)
   a. inclined plane.   c. movable pulley.
   b. wheel and axle.   d. fixed pulley.
9. A gear in a watch is an example of a
   a. pulley.   c. lever.
   b. wheel and axle.   d. screw.
10. An example of a compound machine is a
   a. school bus.   c. crowbar.
   b. pliers.   d. ramp.

True or False

If the statement is true, write "true." If it is false, change the underlined word or words to make the statement true.

1. Work equals force times <u>time</u>.
2. The unit of work in the <u>metric</u> system is the newton-meter or <u>joule</u>.
3. Power is work divided by <u>force</u>.
4. In the metric system, the unit of power is the <u>newton</u>.
5. Work done by a machine is called <u>work output</u>.
6. <u>Friction</u> reduces the efficiency of a machine.
7. A <u>pulley</u> is an inclined plane that moves.
8. An ax is an example of a <u>wheel and axle</u>.

Concept Mapping

Complete the following concept map for Section 4–1. Refer to pages S6–S7 to construct a concept map for the entire chapter.

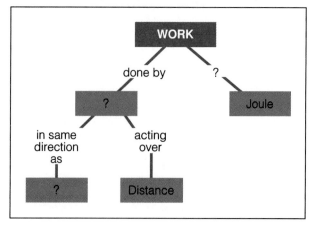

Concept Mastery

Discuss each of the following in a brief paragraph.

1. The mythical god Atlas is known for the fact that he holds up a stationary Earth. Does Atlas perform any work? Explain your answer.
2. Explain how machines make work easier. Use several examples in your answer.
3. Why is a dull razor less helpful than a sharp razor?
4. What is the difference between a screw whose threads are close together and a screw whose threads are far apart?
5. You have been given the task of moving a huge box of baseballs to a shelf that is 1 meter above the floor. There are three ramps available for you to use. They all have a height of 1 meter. However, the first one is 1 meter long, the second is 2 meters long, and the third is 3 meters long. Which ramp will make your task easiest? Why?
6. How can you increase the efficiency of a pulley?
7. Why is a bicycle a compound machine?

Critical Thinking and Problem Solving

Use the skills you have developed in this chapter to answer each of the following.

1. **Applying information** You are walking along the road when you stop to help someone who is changing a tire. Using a jack like the one shown on the right, you lift the rear end of the car with one hand. What type of simple machine is a car jack? How does it work? Does a car jack multiply force or distance? How does the length of the arm affect the use of the jack?

2. **Applying definitions** For each of the following situations, determine whether work is being done. Explain each answer.
   a. You are babysitting for a friend by watching the child while it naps.
   b. You are doing homework by reading this chapter.
   c. You are doing homework by writing answers to these questions.

3. **Developing a model** Suppose you have a large crate you wish to lift off the floor. To accomplish this you have been given a pulley and some rope. The crate has a hook on it, as does the ceiling. Describe two ways in which you could use this

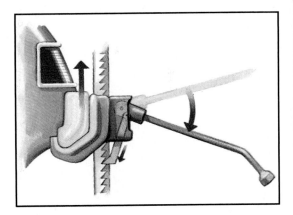

equipment to raise the crate. Accompany each description with a diagram showing how you would set it up.

4. **Using the writing process** You are a brilliant and creative inventor famous for your unusual machines. You have recently completed your most outstanding project—an odd-looking, but very important machine. Write an explanation that will be given to the scientific world describing your machine, how you built it, what it is composed of, and what it does.

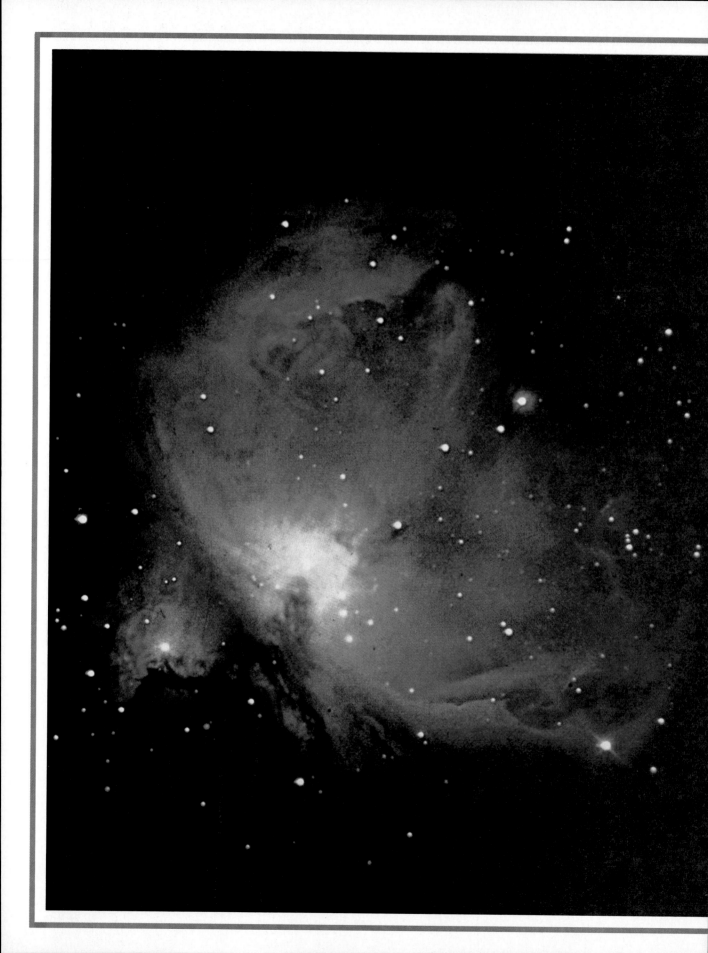

Energy: Forms and Changes

Guide for Reading

After you read the following sections, you will be able to

5–1 Nature of Energy
- ■ Identify five forms of energy.

5–2 Kinetic and Potential Energy
- ■ Compare kinetic energy and potential energy.
- ■ Relate kinetic energy to mass and velocity.

5–3 Energy Conversions
- ■ Describe different types of energy conversions.

5–4 Conservation of Energy
- ■ Explain what Einstein said about the relationship between matter and energy.

5–5 Physics and Energy
- ■ Relate the law of conservation of energy to motion and machines.

Within the large, cold clouds of gas and dust, small particles begin to clump together. Their own gravitational force and the pressure from nearby stars cause these small clumps to form a single large mass. Like a monstrous vacuum cleaner, gravitational force attracts more and more particles of dust and gas. The gravitational force becomes so enormous that the bits of matter falling faster and faster to the center begin to heat up. The internal temperature reaches 15 million degrees. Subatomic particles called protons collide with one another at tremendous speeds. The normal electromagnetic force of repulsion between protons is overcome by the force of particle collisions. The protons fuse together to form helium. During this process, part of the matter is transformed into energy. A star is born!

What is energy? How can energy from our sun be changed into useful energy on the Earth? Is the total energy in the universe constant? As you read further, you will find answers to these questions.

Journal *Activity*

You and Your World Think about the last time you were awakened by a violent thunderstorm. Were you frightened or excited? Did you pay more attention to the lightning or to the thunder? What did you think might happen? In your journal, explore the ideas and feelings you had on this occasion. Include any questions you have about thunderstorms.

◀ *The formation of the Orion Nebula is evidence of the interaction of matter and energy.*

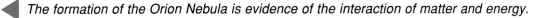

Guide for Reading

*Focus on these questions as
you read.*

▶ *What is energy?*

▶ *What are the different forms
of energy?*

5–1 Nature of Energy

On July 4, 1054, the sudden appearance of a new star was recorded by the Chinese. The star shone so brightly that it could be seen even during the day. After 23 days, the distant star began to disappear. What the Chinese had observed was an exploding star, or supernova. The energy released by a supernova is capable of destroying a nearby solar system in just a few hours. A supernova is one of the greatest concentrations of energy in the universe.

A supernova is a very dramatic example of energy release. But not all forms of energy are quite that dramatic. In fact, you live in an ocean of **energy.** Energy is all around you. You can hear energy as sound, you can see it as light, and you can feel it as wind. You use energy when you hit a tennis ball, compress a spring, lift a grocery bag. Living organisms need energy for growth and movement. Energy is involved when a bird flies, a bomb explodes, rain falls from the sky, and electricity flows in a wire.

Figure 5–1 *Energy is all around you, continually shaping and reshaping the Earth and maintaining all the life that exists upon it.*

What is energy that it can be involved in so many different activities? **Energy can be defined as the ability to do work.**

If an object or organism does work (exerts a force over a distance to move an object), the object or organism uses energy. You use energy when you swim in a race. Electric charges in a current use energy as they move along a wire. A car uses energy to carry passengers from one place to another. Because of the direct connection between work and energy, energy is measured in the same unit as work. Energy is measured in joules (J).

In addition to using energy to do work, objects can gain energy because work is being done on them. If work is done on an object, energy is given to the object. When you kick a football, you give some of your energy to the football to make it move. When you throw a bowling ball, you give it energy. When that bowling ball hits the pins, it loses some of its energy to the pins, causing them to fall down.

Figure 5–2 *Although he may not realize it, this young boy has energy simply because he is in motion. The same is true of the cascading waterfall.*

Forms of Energy

Energy appears in many forms. **The five main forms of energy are mechanical, heat, chemical, electromagnetic, and nuclear.** It may surprise you to learn that your body is an "energy factory" that stores and converts various forms of energy. After reading about each form of energy, see if you can describe how your energy factory works.

MECHANICAL ENERGY Matter that is in motion has energy. The energy associated with motion is called **mechanical energy.** Water in a waterfall has a great amount of mechanical energy. So does wind. An automobile traveling at 95 km/hr has mechanical energy. A jet plane cruising at 700 km/hr has even more! When you walk, ride a bike, or hit a ball, you use mechanical energy. Sound is a type of mechanical energy. Even the blood flowing through your blood vessels has mechanical energy.

HEAT ENERGY All matter is made up of tiny particles called atoms that are constantly moving. The internal motion of the atoms is called **heat energy**. The faster the particles move, the more heat energy is produced. Rub your hands together for several

ACTIVITY

DOING

Energy in the News

1. Make five columns on a sheet of paper.

2. Write one of the following headings at the top of each column: Mechanical Energy, Heat Energy, Chemical Energy, Electromagnetic Energy, and Nuclear Energy.

3. Read through a newspaper and place a check mark in the appropriate column every time a particular form of energy is mentioned.

What form of energy is most often mentioned? Would this form of energy always be the most discussed?

Figure 5–3 *Lots of delicious foods to eat have an added benefit. They are a source of energy. What type of energy is stored in food?*

seconds. Did you feel heat? Using the friction between your hands, you converted mechanical energy (energy of motion) into heat energy! Heat energy usually results from friction. Heat energy causes changes in the temperature and phase (solid, liquid, gas) of any form of matter. For example, it is heat energy that causes your ice cream cone to melt and drip down your hand.

CHEMICAL ENERGY Energy is required to bond atoms together. This energy is called **chemical energy**. Often, when bonds are broken, this chemical energy is released. The fuel in a rocket engine has stored chemical energy. When the fuel is burned, chemical energy is released and converted into heat energy. When you start a fire in a charcoal grill, you are releasing chemical energy. When you digest food, bonds are broken to release energy for your body to store and use. When you play field hockey or lacrosse, you are using the chemical energy stored in your muscles that you obtained from food.

ELECTROMAGNETIC ENERGY Moving electric charges have the ability to do work because they have **electromagnetic energy.** Power lines carry electromagnetic energy into your home in the form of electricity. Electric motors are driven by electromagnetic energy. Light is another form of electromagnetic energy. Each color of light—red, orange, yellow, green, blue, violet—represents a different amount of electromagnetic energy. Electromagnetic energy is also carried by X-rays, radio waves, and laser light.

NUCLEAR ENERGY The nucleus, or center, of an atom is the source of **nuclear energy.** When the nucleus splits, nuclear energy is released in the form of heat energy and light energy. Nuclear energy is also released when lightweight nuclei collide at high speeds and fuse (join). The sun's energy is produced from a nuclear fusion reaction in which hydrogen nuclei fuse to form helium nuclei. Nuclear energy is the most concentrated form of energy.

Figure 5–4 *Light, whether seen as a beautiful rainbow or used as laser beams, is an important part of everyday life. No matter how it is used, light is a form of energy. What form of energy is light?*

5–1 Section Review

1. What is energy?
2. Can energy be transferred from one object to another? Explain.
3. What are the different forms of energy?
4. Why is energy measured in the same unit as work?

Connection—*You and Your World*

5. It is energy you must pay for on your electric bill. Electric companies usually express the total amount of energy used in kilowatt-hours (kW-h)—the flow of 1 kilowatt of electricity for 1 hour. How many joules of energy do you get when you pay for 1 kW-h? (1 J = 1 watt x 1 second; 1 kW = 1000 watts; 1 hour = 3600 seconds)

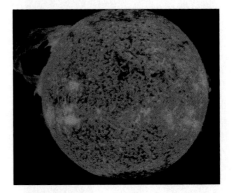

Figure 5–5 *A solar flare erupts from the sun at temperatures greater than 50,000°C. What form of energy is responsible for the characteristics of the sun?*

5–2 Kinetic and Potential Energy

Guide for Reading

Focus on this question as you read.

▶ *What is the difference between kinetic energy and potential energy?*

Stretch a rubber band between your thumb and index finger. Keep the rubber band stretched without any motion. How long can you hold it this way? After a short while, as your fingers begin to tire, you become aware of the energy in the rubber band. Yet the rubber band is not moving! This is because the energy of the stretched rubber band is stored in it. Remember that energy is the ability to do work. Release your thumb and the rubber band moves. As the rubber band moves back to its normal shape, it does work.

The energy that you felt when you stretched the rubber band was different from the energy displayed when the rubber band snapped back to its original shape. They are two different states of energy. The five different forms of energy you just learned about can be classified into either one of these states of energy. The two states are called **kinetic energy** and **potential energy.**

Figure 5–6 *One powerful bang and the energy in a moving hammer enables this youngster to win a prize. The same type of energy—energy of motion—can demolish a building. What is this type of energy called?*

Figure 5–7 *Although this horse and colt may be running at the same velocity, they each have a different amount of kinetic energy because they have different masses.*

Kinetic Energy

An object that is moving can do work on another object by colliding with that object and moving it through a distance. A flying rubber band does work when it flattens a house of cards. A swinging hammer does work on a nail as it drives the nail into a piece of wood. A wrecking ball does work as it knocks down a wall. Because an object in motion has the ability to do work, it has energy. **The energy of motion is called kinetic energy.** The word kinetic comes from the Greek word *kinetikos* which means "motion." Why do the particles in matter have kinetic energy?

Suppose you are accidentally hit with a tennis ball that has been tossed lightly toward you. It probably does not hurt you. Now suppose you are hit with the same tennis ball traveling at a much greater speed. You can certainly feel the difference! The faster an object moves, the more kinetic energy it has. So kinetic energy is directly related to the velocity of an object. You have more kinetic energy when you run than when you walk. In baseball, a fast ball has more kinetic energy than a slow curve. When does a skier have more kinetic energy, when skiing downhill or cross-country?

Do all objects with the same velocity have the same kinetic energy? Think about the tennis ball again. Suppose this time it rolls across the tennis court and hits you in the foot. Compare this with getting hit in the foot by a bowling ball traveling at the same speed as the tennis ball. The bowling ball is much more noticeable because the bowling ball has more kinetic energy than the tennis ball. A battleship moving at 40 km/hr has much more kinetic energy than a mosquito moving at the same velocity. So kinetic energy must depend on something other than just velocity. The battleship has more kinetic energy because it has greater mass. Kinetic energy depends on both mass and velocity. The mathematical relationship between kinetic energy (K.E.), mass, and velocity is:

$$\text{K.E.} = \frac{\text{mass} \times \text{velocity}^2}{2}$$

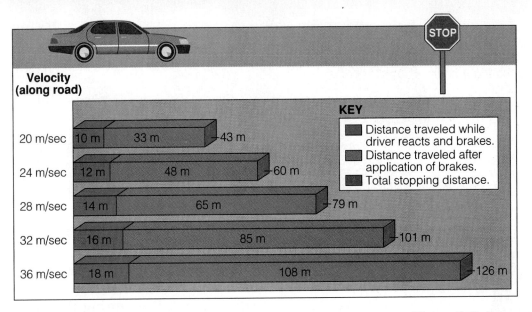

Velocity (along road)

Velocity	Distance traveled while driver reacts and brakes	Distance traveled after application of brakes	Total stopping distance
20 m/sec	10 m	33 m	43 m
24 m/sec	12 m	48 m	60 m
28 m/sec	14 m	65 m	79 m
32 m/sec	16 m	85 m	101 m
36 m/sec	18 m	108 m	126 m

KEY
- Distance traveled while driver reacts and brakes.
- Distance traveled after application of brakes.
- Total stopping distance.

Figure 5–8 *A car requires a longer distance in which to stop when traveling at faster velocities. Notice how quickly the distance increases for a small increase in velocity.*

According to this equation, an increase in either mass or velocity will mean an increase in kinetic energy. Which of these two factors, mass or velocity, will have a greater effect on kinetic energy? Why?

Now suppose you want to push a heavy box across the floor. You must exert a force on the box to move it. Thus you do work on the box. Before you moved the box, it did not have kinetic energy because it did not have velocity. As you give it kinetic energy, the box picks up velocity. The more work you do, the faster the box will move. When you increase the velocity of an object, you increase its kinetic energy. The change in the kinetic energy of the box is equal to the work you have done on it.

Potential Energy

You just read that some objects are able to do work as a result of their motion. Other objects can do work because of their position or shape. **Potential energy is energy of position.** A stretched rubber band has the potential, or ability, to fly across the room. A wound-up watch spring also has potential energy. It has the potential to move the hands of the watch around when it unwinds. An archer's taut (tightly stretched) bow has the potential to send an arrow gliding toward a target. A brick being held high above the ground has the potential to drive a stake into the ground when it falls onto it.

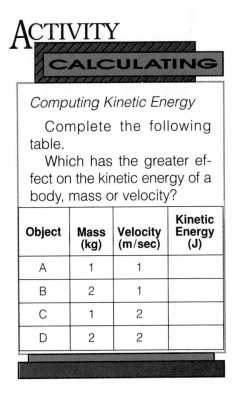

ACTIVITY

CALCULATING

Computing Kinetic Energy

Complete the following table.

Which has the greater effect on the kinetic energy of a body, mass or velocity?

Object	Mass (kg)	Velocity (m/sec)	Kinetic Energy (J)
A	1	1	
B	2	1	
C	1	2	
D	2	2	

Figure 5–9 *A jack-in-the-box uses potential energy to burst out of its container. How does the archer use potential energy?*

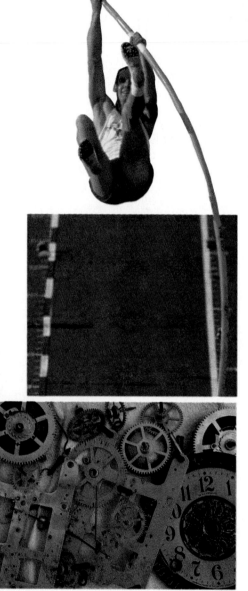

Figure 5–10 *Tightly wound springs store potential energy that can be used to turn the hands of time. What kind of potential energy does a pole vaulter have at the top of a vault?*

Potential energy is related to work in a different way than kinetic energy is. Remember that a moving object has kinetic energy because it can do work as it moves. But an object with potential energy is not moving or doing work. Instead, it is storing the energy that was given to it when work was done on it. It has the ability, or potential, to give that energy back by doing work. The spring acquired potential energy because work was done on it by the person winding the watch. Work was done by the person who pulled back on the bow's arrow. The brick acquired potential energy because work was done in lifting it.

Potential energy is not always mechanical, or associated with movement. For example, the chemical energy stored in food is an example of potential energy. The energy is released when the food is broken down in digestion and respiration. Similarly, fuels such as coal and oil store chemical potential energy. The energy is released when the fuel is burned. The nucleus of an atom consists of a number of particles held together by a strong force. The potential energy stored in the nucleus of an atom can be released if the nucleus is split in a nuclear reactor.

GRAVITATIONAL POTENTIAL ENERGY Imagine that you are standing on the edge of a 1-meter diving board. Do you think you have any energy? You probably think you do not because you are not moving. It is true that you do not have kinetic energy. But you do have potential energy. Your potential energy is due to your position above the water.

If you stand on a 3-meter diving board, you have three times the potential energy you have on the 1-meter board. Potential energy that is dependent on height is called **gravitational potential energy.** A waterfall, suspension bridge, and falling snowflake all have gravitational potential energy.

Weight also determines the amount of gravitational potential energy an object has. The old saying "The bigger they are, the harder they fall" is an observation of the effect of weight on gravitational potential energy. From your experiences, you may already know that gravitational potential energy is dependent on weight. You have a lot more gravitational potential energy with a heavy pack on your back than you do with a light pack.

The relationship between gravitational potential energy (G.P.E.), weight, and height can be expressed by the following formula:

$$\textbf{G.P.E. = Weight} \times \textbf{Height}$$

You can see from this formula that the greater the weight, the greater the gravitational potential energy. The higher the position above a surface, the greater the gravitational potential energy.

Figure 5–11 *This huge boulder in Arches National Park in Utah has a great deal of gravitational potential energy. So does a falling drop of water in a leaky faucet. The drop of water could not fall without the help of energy. Rock climbers, on the other hand, must do a tremendous amount of work to increase their gravitational potential energy.*

5–2 Section Review

1. What is kinetic energy? Potential energy?
2. Use the formula for kinetic energy to describe the relationship between the kinetic energy of an object, its mass, and its velocity.
3. What is gravitational potential energy? How is it calculated?

Critical Thinking—*Relating Concepts*
4. If you use the sun as your frame of reference, you always have kinetic energy. Why?

Figure 5–12 *As a basketball player throws the ball in the air, various energy conversions take place. What are these conversions?*

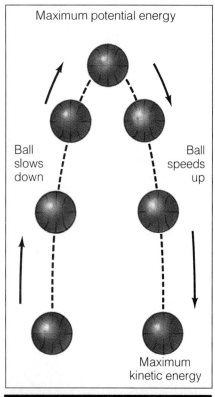

Maximum potential energy

Ball slows down

Ball speeds up

Maximum kinetic energy

5–3 Energy Conversions

When you think of useful energy, you may think of the energy involved in moving a car or the energy you get from the food you eat. But in both these examples, the useful energy was obtained by first converting energy from one form to another. The mechanical energy of the car came from burning the chemical energy of fuel. Your energy also comes from chemical energy—the chemical energy stored in the food you eat. Energy can be transferred from one object to another and energy can be changed from one form to another. Changes in the forms of energy are called **energy conversions.**

Kinetic–Potential Energy Conversions

One of the most common energy conversions involves the changing of potential energy to kinetic energy or kinetic energy to potential energy. A stone held high in the air has potential energy. As it falls, it loses potential energy because its height decreases. At the same time its kinetic energy increases because its velocity increases. Thus potential energy is converted into kinetic energy. Similarly, the potential energy stored in a bent bow can be converted into the kinetic energy of the arrow.

Conversions between kinetic energy and potential energy are taking place around you every day. Think of tossing a ball up into the air. When you throw the ball up, you give it kinetic energy. As the ball rises, it slows down. As its velocity decreases, its kinetic energy is reduced. But at the same time its height above the Earth is increasing. Thus its potential energy is increasing. At the top of its path, the ball has slowed down to zero velocity so it has zero kinetic energy. All of its kinetic energy from the beginning of its flight has been converted to potential energy.

Then the ball begins to fall. As it gets closer to the Earth's surface, its potential energy decreases. But it is speeding up at the same time. Thus its kinetic energy is increasing. When you catch it, it has its maximum velocity and kinetic energy. The potential energy of the ball has changed into kinetic energy.

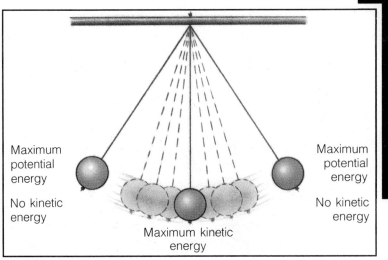

Figure 5–13 *A continuous conversion between kinetic energy and potential energy takes place in a pendulum. Potential energy is greatest at the two highest points in the swing and zero at the bottom. Where is kinetic energy greatest?*

Labels in figure:
- Maximum potential energy / No kinetic energy
- Maximum potential energy / No kinetic energy
- Maximum kinetic energy

Other Conversions

Although conversions between kinetic energy and potential energy are common, they are not the only changes in energy that take place. **All forms of energy can be converted to other forms.** For example, the sun's energy is not used merely as heat energy or light energy. It is converted to other forms of energy as well. Solar products convert the energy of sunlight directly into electricity. Green plants use the energy of the sun to trigger a process in which sugars and starches are made. These substances store the energy as chemical energy. In this process, electromagnetic energy is converted to chemical energy.

In an electric motor, electromagnetic energy is converted to mechanical energy. In a battery, chemical energy is converted to electromagnetic energy. The mechanical energy of a waterfall is converted to electromagnetic energy in a generator. Solar cells convert the sun's energy directly into electrical energy. In a heat engine (such as an automobile engine), fuel is burned to convert chemical energy into heat energy. The heat energy is then changed to mechanical energy. In a microphone-loudspeaker system, the microphone converts mechanical energy in the form of sound into electromagnetic energy in the form of electricity. The electromagnetic energy goes to the loudspeaker that then converts the electric signal back into sound.

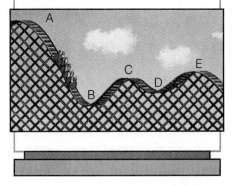

Power plant

Chemical
energy

Heat
energy

Mechanical
energy

Electric
energy

Heat
energy

Figure 5–14 *A series of energy conversions is needed to produce the heat energy of the hair dryer. Trace the conversions.*

ACTIVITY

Mixing It Up

1. Fill two mixing bowls with cold water. Record the temperature of the water in each bowl.

2. Run an electric or hand mixer in one bowl for a few minutes.

3. Take the temperature of the water in each bowl again. You must read the thermometer with great precision.

How did you expect the temperatures to compare after you used the mixer?

How do the final temperatures actually compare? Why?

Often a whole series of energy conversions is needed to do a particular job. The operation of a hair dryer is a good example of this. See Figure 5–14. The electromagnetic energy used by the dryer is generated from some fuel source, such as gas. The chemical energy of the fuel is released by burning it. The fuel provides heat energy, which in turn is changed to mechanical energy. This mechanical energy is used to make a generator do the work of providing the dryer with electromagnetic energy in the form of electricity. When you turn the dryer on, the electricity is changed back to heat energy.

Figure 5–15 *This cute little animal is enjoying an afternoon snack totally unaware that the complex process of energy conversion is taking place. Describe the energy conversions.*

5–3 Section Review

1. Describe the conversions between potential energy and kinetic energy as a tennis ball drops, hits the ground, and bounces back up.
2. What energy conversions take place in a pendulum? Why does the pendulum eventually stop?
3. Describe the energy conversions that you think take place when a flashlight is turned on.

Critical Thinking—*Analyzing Information*

4. Identify the various energy conversions involved in the following events: An object is raised and then allowed to fall. As it hits the ground it stops, produces a sound, and becomes warmer.

Activity Bank

Crazy Eights, p.151

5–4 Conservation of Energy

When you turn on a lamp, not all of the electricity flowing through the filament of the light bulb is converted into light. This may lead you to think that energy is lost. But it is not. It is converted into heat. Although heat is not useful in a lamp, it is still a form of energy. Energy is never lost. Scientists have found that even when energy is converted from one form to another, no energy is gained or lost in the process. **The law of conservation of energy states that energy can be neither created nor destroyed by ordinary means.** Energy can only be converted from one form to another. So energy conversions occur without a loss or gain in energy.

The **law of conservation of energy** is one of the foundations of scientific thought. If energy seems to disappear, then scientists look for it. Important discoveries have been made because scientists believed so strongly in the conservation of energy.

One such discovery was made by Albert Einstein in 1905. Part of his famous theory of relativity deals with the concept that mass and energy are directly related. According to Einstein, even the tiniest mass can form a tremendous amount of energy. With this mass-energy relationship, Einstein was saying that mass and energy can be converted into each other.

Guide for Reading

Focus on this question as you read.

▶ *What is the law of conservation of energy?*

Figure 5–16 *You have probably felt the heat given off by a light bulb. The heat released is energy that did not become light.*

With this relationship, Einstein clarified the law of conservation of energy. He showed that if matter is destroyed, energy is created and if energy is destroyed, matter is created. The total amount of mass and energy is conserved.

PROBLEM Solving

How Energy Conscious Are You?

Years ago, most people believed that the Earth's energy resources were endless. People used energy at an astonishing rate with no concern that the resources might someday run out. Today, ideas are changing. People have begun to face the reality that we can no longer afford to waste these precious resources. Until practical alternatives are found, we must make every effort to conserve those resources we have. Answer the following questions to find out how energy conscious you are.

1. If you leave a room for at least an hour, do you leave the electrical appliances in the room on—lights, television, radio?

2. When you want something from the refrigerator, do you stare into the open refrigerator while you slowly decide what you want?

3. Suppose you are running in and out of a room every few minutes. Do you turn the light on and off every time you walk into or out of the room?

4. Do you open and close the oven to peek at the brownies cooking?

5. Do you wait for a ride in a car or bus rather than riding your bike or walking to nearby locations?

6. Do you take the elevator instead of the stairs?

7. Do you throw away bottles and cans rather than saving them for recycling?

If you answered "no" to all or most of these questions, you are in good shape. If not, you may be hazardous to the environment! And what's more, your monthly electric bill reflects the energy you use. If you save energy, you save money.

Relating Cause and Effect

Think about the questions to which you answered "no." How do your actions conserve energy? Think about the questions to which you answered "yes." Why are your actions not energy efficient? Make a list of ways in which you could start saving energy every day. Think of as many additional examples as you can.

During nuclear reactions—such as those that take place in the sun—energy and mass do not seem to be conserved. But Einstein showed that a loss in mass results in a gain in energy. Mass is continuously changed to energy in our sun through a process called nuclear fusion. During this process, a small loss in mass produces a huge amount of energy.

Figure 5–17 *In 1905, Albert Einstein (1879–1955) made a major contribution to science with his theory of relativity. Part of the theory describes a direct relationship between mass and energy.*

5–4 Section Review

1. What is the law of conservation of energy? How does it relate to energy conversions?

Critical Thinking—*Making Inferences*
2. Using the law of conservation of energy, explain why you become tired from pushing your bicycle along the road.

5–5 Physics and Energy

The topic of energy is essential to learning about any subject in physical science. You may wonder, then, why you have not studied energy earlier. Would it surprise you to know that you have actually been learning about energy for the past several chapters? For example, you learned that you travel at a faster *speed* when you run than when you walk. Now it should be clear that you can travel at a faster speed if you use more energy. You must exert more energy to pedal your bicycle quickly than to ride along slowly.

You learned that a moving object, such as a billiard ball, has *momentum* (mass x velocity.) The quantities mass and velocity are also used to measure the kinetic energy of an object. Thus an object that has momentum also has kinetic energy. Momentum must be conserved because energy is conserved. A moving billiard ball that collides with a stationary one gives some or all of its energy to the other ball, causing it to move.

You learned that a *force* is required to change the motion of an object. A force acting on an object to change its motion is doing work. If a force does

Focus on this question as you read.

▶ *How is the law of conservation of energy related to other physical principles?*

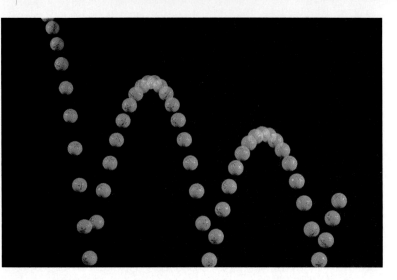

Figure 5–18 *If you follow the bouncing ball, you will see that it gets lower and lower. The forces of friction and gravity are responsible for this behavior. Is the energy of the ball lost?*

work on an object, it changes the energy of the object. When a ball is bounced off the head of a soccer player, the player exerts a force on the ball. In so doing, the player gives energy to the ball, causing a change in its motion.

You learned that *power* is the rate at which work is done. Thus power must be the rate at which energy is consumed. When you run instead of walk up the stairs, you use the same amount of energy (do the same amount of work), but you use the energy at a faster rate. Your monthly electric bill measures the electromagnetic energy you use. The electric company multiplies the power you used by the length of time it was used.

Another quantity you learned about is *work.* Now you know that work is directly related to energy. When you read about machines, you also learned that the work that comes out of a machine can never be greater than the work that goes into a machine. Work done on a machine means that energy goes into the machine. Because energy is conserved, the same amount of energy must come out of the machine. Thus, since energy is conserved, work must also be conserved. The only energy that does not come directly out of a machine is that taken by friction. But this energy is not lost, it is simply converted to another form—heat.

Have you begun to see that almost nothing happens without the involvement of energy? It is interesting to note that the concept of energy was not yet developed in Newton's time. However, once energy was described, Newton's detailed descriptions of motion could be easily explained in terms of energy. In addition, the laws of motion did not violate the law of conservation of energy. In fact, no physical phenomena have yet violated the law of conservation of energy.

Follow the Bouncing Ball

1. Hold a meterstick vertically, with one end on the ground.

2. Drop a tennis ball from the 50-centimeter mark and record the height to which it rebounds.

3. Drop the tennis ball from the 100-centimeter mark and record the height of the rebound.

■ What can you conclude about gravitational potential energy and height?

Figure 5–19 *This art by M. C. Escher shows an unusual waterfall that violates the law of conservation of energy. When the water falls, part of its potential energy is converted into the kinetic energy of the water wheel. But how does the water get back up to the top?*

5–5 Section Review

1. How is energy related to motion?
2. How is energy related to force?
3. How is energy related to power?

Critical Thinking—*Making Calculations*

4. Two cars have the same momentum. One car weighs 5000 N and the other weighs 10,000 N. Which car has a greater kinetic energy? Explain your answer.

CONNECTIONS

Our Energetic World

Start looking around. Imagine that you are standing on the edge of the Grand Canyon in Arizona. In the distance you see a wonderful snowcapped mountain.

Mountains result from great forces that push up through the Earth's surface. Think about the tremendous amount of work, and therefore energy, that was required to create the mountain. The huge amount of energy located within the Earth's interior not only creates mountains, it also causes volcanic eruptions, earthquakes, and movement of the continents. There, you have seen the effect of energy already without too much trouble.

Look some more—this time downward. Think about the amount of work required to chisel out such a masterpiece as the Grand Canyon. Clearly, a great deal of energy was involved. The Grand Canyon was dug out by the Colorado River over millions of years. Water and other natural forces continually reshape the face of the *Earth*. Glaciers, ocean waves, winds, and rough storms are examples of the natural forces that not only create some of the most beautiful sights in the world,

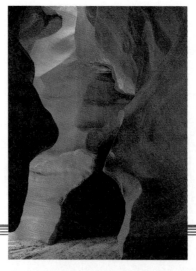

but also can be some of the most destructive forces on Earth.

Perhaps all this thinking is making you hungry. Why not sit down and eat lunch. Guess what! Energy is there again. It's in

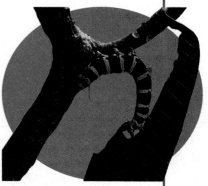

the food you eat. Food, like any fuel, has chemical potential energy stored in it. When you eat it, your digestive and respiratory systems break down the food and release energy into your system. You use this energy to keep all your body systems working and to power your daily activities. You would not be able to walk around, run, play, or even think without the energy you obtain from the food you eat.

Now you might wonder where the food you eat gets its energy. The sun is the ultimate source of energy for all *living things*. Green plants and certain bacteria trap the energy of sunlight. They use about half of this energy for their own activities and store the rest in compounds that they manufacture called carbohydrates. Animals that either eat the plants or eat other animals that eat plants obtain this stored energy.

Wow! You can't seem to get away from energy. Take a deep breath and relax. Oh, there it is again. The oxygen in the air you breathe is released in the

process by which plants convert the energy of the sun into the chemical energy of food. So without the energy that is required to drive these conversion processes, there would not be any oxygen for you to breathe.

Maybe you should walk around a bit and organize your thoughts. Surprise! Energy is involved once again. When you walk, you convert chemical energy in your body to mechanical energy and heat. In fact, you use mechanical energy for your movements, chemical energy in your body processes, and electrical energy to control many of your body systems. And when you move something by throwing it, pushing it, or picking it up, you give some of your energy to the object, causing it to move.

Boy it's sure getting hot out there thinking about energy. Speaking of heat leads us to the sun. The sun is the main source of energy for living things. Without the sun, there would be no life on Earth. The sun releases energy in a process called nuclear fusion. Within the core of the sun, ex-

tremely strong gravitational forces pull the atoms of hydrogen gas together so tightly that they fuse into helium atoms. During this process, some mass is changed to energy, mostly in the form of heat and light. A portion of this energy reaches the Earth.

The sun is only one of the trillions and trillions of stars in the universe. Nonetheless, the sun is the center of our solar system. It showers the Earth and the eight other planets with a constant supply of energy. And as you have been learning, thanks to the sun, the Earth is alive with energy.

Have you begun to see things a little differently? Energy in one form or another is everywhere and is necessary for every activity and process you can imagine. From the cry of a baby to the falling of rain to the rumble of an earthquake, energy is involved. You use energy constantly—even when you are asleep. In fact, because the Earth moves around the sun, you and everything on the Earth always have kinetic energy. So the next time you need a little energy, just look around!

Laboratory Investigation

Relating Mass, Velocity, and Kinetic Energy

Problem

How does a change in mass affect the velocity of an object if its kinetic energy is constant?

Materials *(per group)*

rubber band
3 thumbtacks
12 washers glued together in groups of 2, 4, and 6
wooden board, 15 cm x 100 cm
meterstick

Procedure 👁 ▆

1. Place three thumbtacks at one end of the wooden board, as shown in the figure.

2. Stretch the rubber band over the three thumbtacks to form a triangle.

3. In front of the rubber band, place two washers that have been stuck together.

4. Pull the washers and the rubber band back about 2 cm, as in the figure. Release the rubber band. The washers should slide about 70 to 80 cm along the board.

5. Practice step 4 until you can make the double washer travel 70 to 80 cm each time.

6. Mark the point to which you pulled the rubber band back to obtain a distance of 70 to 80 cm. This will be your launching point for the entire experiment.

7. Launch the double washer three times. In a data table, record the distance in centimeters for each trial. Remember to use the same launching point each time.

8. Repeat step 7 for a stack of 4 washers.

9. Repeat step 7 for a stack of 6 washers.

Observations

Calculate the average distance traveled by 2 washers, 4 washers, and 6 washers.

Analysis and Conclusions

1. What is the relationship between the mass, or number of washers, and the average distance traveled?

2. What kind of energy was in the washers when you held them at the launching point? How do you know?

3. After the washers were launched, what kind of energy did they have?

4. You launched all the washers from the same position. Was the energy the same for each launch?

5. Assume that the farther the washers slid, the greater their initial velocity. Did the heavier group of washers move faster or slower than the lighter group?

6. If the kinetic energy is the same for each set of washers, what happens to the velocity as the mass increases?

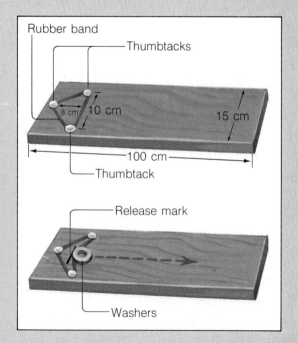

Study Guide

Summarizing Key Concepts

5–1 Nature of Energy

▲ Energy is the ability to do work.

▲ Energy appears in many forms: mechanical, heat, chemical, electromagnetic, and nuclear.

5–2 Kinetic and Potential Energy

▲ Energy that an object has due to its motion is called kinetic energy.

▲ Kinetic energy equals one half the product of the mass times the square of the velocity.

▲ Energy that an object has due to its shape or position is called potential energy.

▲ Potential energy that an object has due to its height above the Earth's surface and its weight is called gravitational potential energy.

5–3 Energy Conversions

▲ Energy can change from one form to another. Changes in the form of energy are called energy conversions.

▲ The most common energy conversions occur between kinetic energy and potential energy. But all forms of energy can be converted to another form.

5–4 Conservation of Energy

▲ The law of conservation of energy states that energy can neither be created nor destroyed by ordinary means.

5–5 Physics and Energy

▲ Energy is involved in every physical activity or process.

▲ An increase in speed or velocity is accompanied by an increase in kinetic energy.

▲ An object that has kinetic energy also has momentum.

▲ A force doing work on an object to change its motion is giving energy to the object.

▲ Power is the rate at which energy is used.

▲ The conservation of work can be understood because energy is conserved.

Reviewing Key Terms

Define each term in a complete sentence.

5–1 Nature of Energy
energy
mechanical energy
heat energy
chemical energy
electromagnetic energy
nuclear energy

5–2 Kinetic and Potential Energy
kinetic energy
potential energy
gravitational potential energy

5–3 Energy Conversions
energy conversion

5–4 Conservation of Energy
Law of Conservation of Energy

Chapter Review

Content Review

Multiple Choice

Choose the letter of the answer that best completes each statement.

1. Energy is the ability to do
   a. motion.   c. acceleration.
   b. work.   d. power.
2. The unit in which energy is measured is the
   a. newton.   c. electron.
   b. watt.   d. joule.
3. X-rays, lasers, and radio waves are forms of
   a. mechanical energy.
   b. heat energy.
   c. electromagnetic energy.
   d. nuclear energy.
4. Gravitational potential energy is dependent on
   a. speed and height.
   b. weight and height.
   c. time and weight.
   d. acceleration and kinetic energy.

5. Gasoline and rocket fuel store
   a. electromagnetic energy.
   b. chemical energy.
   c. mechanical energy.
   d. gravitational potential energy.
6. A stretched rubber band has
   a. potential energy.
   b. kinetic energy.
   c. nuclear energy.
   d. electromagnetic energy.
7. Energy of motion is
   a. potential energy.
   b. nuclear energy.
   c. kinetic energy.
   d. electromagnetic energy.
8. According to Einstein, matter is another form of
   a. mass.   c. time.
   b. light.   d. energy.

True or False

If the statement is true, write "true." If it is false, change the underlined word or words to make the statement true.

1. Energy is the ability to do <u>work</u>.
2. The food you eat stores <u>chemical</u> energy.
3. Light is <u>nuclear</u> energy.
4. Sound is a form of <u>mechanical</u> energy.
5. Energy stored in an object due to its position is called <u>kinetic</u> energy.
6. <u>Potential</u> energy is energy of motion.
7. Kinetic energy equals the mass of an object times the square of its velocity divided by <u>two</u>.
8. Gravitational potential energy is dependent on both the weight and <u>height</u> of an object.

Concept Mapping

Complete the following concept map for Section 5–1. Refer to pages S6–S7 to construct a concept map for the entire chapter.

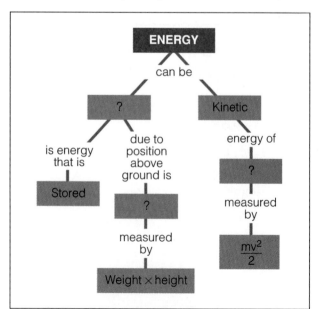

Concept Mastery

Discuss each of the following in a brief paragraph.

1. Describe five different examples of mechanical energy.
2. From the standpoint of kinetic energy, why is a loaded truck more dangerous than a small car in a collision even though they are traveling at the same speed?
3. How does bouncing on a trampoline illustrate both kinetic and potential energies?
4. Water is boiled. The resulting steam is blown against huge turbine blades. The turning blades spin in a magnetic field, producing electricity. Describe in order the energy conversions.
5. How does the law of conservation of energy relate to the following situations: a bat hitting a baseball, a person throwing a frisbee, a person breaking a twig over his or her knee.
6. The concept of energy links the various scientific disciplines—physical science, life science, and earth science. Explain why.

Critical Thinking and Problem Solving

Use the skills you have developed in this chapter to answer each of the following.

1. **Applying concepts** Sound is produced by vibrations in a medium such as air. The particles of air are first pushed together and then pulled apart. Why is sound considered a form of mechanical energy?
2. **Relating concepts** A bear in a zoo lies sleeping on a ledge. A visitor comments: "Look at that lazy bear. It has no energy at all." Do you agree? Explain your answer.
3. **Making calculations and graphs** The gravitational potential energy of a boulder at 100 m is 1000 J. What is the G.P.E. at 50 m? At 20 m? At 1 m? At 0 m? Make a graph of height versus energy. What is the shape of your graph?
4. **Applying concepts** Two cyclists are riding their bikes up a steep hill. Jill rides her bike straight up the hill. Jack rides the bike up the hill in a zigzag formation. Jack and Jill have identical masses. At the top of the hill, does Jack have less gravitational potential energy than Jill? Explain your answer.
5. **Identifying relationships** The diagram shows a golfer in various stages of her swing. Compare the kinetic and potential energies of the golf club at each labeled point in the complete golf swing.
6. **Using the writing process** Imagine that the Earth's resources of coal and oil were suddenly used up. Describe how your typical day would change from morning until night. Give details about what you and your family would do about such things as cooking, transportation, entertainment, heat, and light. Discuss the importance of finding alternative energy resources.

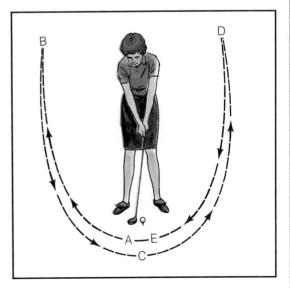

Guion Bluford:

CHALLENGER
in Space

Have you ever been told by someone that you could not do something—only to prove that you could? That is what happened to Guion (GIGH-on) Bluford. Guy had always dreamed of becoming an aerospace engineer. As a boy, he built model airplanes and read his father's engineering books. He analyzed the game of table tennis to find out how many different ways he could hit a Ping-Pong ball in order to alter its flight. Yet when Guy reached his last year in high school, his guidance counselors told him that he was "not college material" and that he should go to a technical school instead. Little did they know that this young man would someday earn a Ph.D. in aerospace engineering and join the NASA space program—and become the first African American to fly in space.

Guy admits, "I really wasn't too concerned about what the counselors said. I just ignored it. I had such a strong interest in aerospace engineering by then that nothing a counselor said was going to stop me."

With his parents' encouragement, Guy applied to Pennsylvania State University and was accepted into the aerospace engineer-

ing program. The courses were tough, but he did well. After receiving a bachelor's degree from Penn State, Guy joined the Air Force and went to Vietnam as a pilot. When he returned to the United States after the war, he was accepted into the Air Force Institute of Technology. There he earned a master's degree and a Ph.D. in aerospace engineering.

Guy decided that NASA would be the best place for him to learn about the latest aerospace technology. In 1978, he applied for a position in the astronaut program. Out of 8878 applicants, only 35 were chosen—and Guy was one of them. He was sent to work at the Johnson Space Flight Center in Houston, Texas.

At the Johnson Space Flight Center, Guy took different courses for the first year. Then he spent the next few years flying in "shuttle simulators"—machines that imitate the look and feel of a space shuttle. The work was so interesting and enjoyable that Guy could not have been happier. In his own words, "The job is so fantastic, I don't need a hobby. My hobby is going to work!"

One day while he was at the Johnson Space Flight Center, Guy received a message that

NASA's "top brass" wanted to see him. Guy's first reaction was to think that he had done something wrong! Much to his amazement, he learned that he had been chosen to fly in the third mission of the *Space Shuttle Challenger*. Guy was so thrilled that he was walking on air–and soon he really would be!

The *Challenger* took off from Cape Canaveral, Florida at 2:00 AM on August 30, 1983. Two of the five astronauts on board were designated as pilots and three were designated as mission specialists. Pilots fly the shuttle, while mission specialists are in charge of scientific experiments. Guy's job was that of mission specialist. In particular, his special task was to send out a satellite for the nation of India.

Guy's training in shuttle simulators had prepared him for the experience of space travel–almost. Once in orbit, Guy found that life in zero gravity took some getting used to.

"There is no feeling of right side up or upside down," he recalls, "When you're floating around in space, you feel the same when you're upside down as you do when you're right side up. You don't feel any different when you're standing on the ceiling than you do when you're standing on the floor. [You can] walk across the ceiling or

along the walls as easily as you can walk on the floor!"

Ordinary tasks such as eating and sleeping become real challenges in a weightless environment. Food has to be something that sticks to a plate, like macaroni and cheese. You cannot eat anything like peas because they will just float away. Knives and forks have to be held to eating trays with little magnets–otherwise they, too, will float away.

Sleeping also presents problems. A weightless astronaut who falls asleep will soon find himself floating all over the spaceship and bumping into things. Some astronauts aboard the *Challenger* slept strapped in their seats, but Guy preferred to tie one end of a string to his waist and the other end to something stable on the walls of the cockpit. Then he would float into the middle of the room and fall asleep. Guy recalls, "Occasionally I'd float up against the lockers and be jarred awake. It felt funny waking up and not knowing if I was upside down or not."

Challenger stayed in orbit for six days, from August 30 to September 5. During that time, the crew received a special telephone message from then President Reagan. Mr. Reagan praised all of the astronauts, but had a special message for Guy Bluford: "You, I think, are paving the way for others, and you are making it plain that we are in an era of brotherhood here in our land."

How does Guion Bluford feel about being the first African American in space? Guy has always stressed that he wants to be known for doing a good job–not for the color of his skin. Yet he is glad to be a role model for others who are African American. He hopes that young people will look at him and say, "If this guy can do it, maybe I can do it too." And he is quick to add that his story has a message for all young people. "They can do it. They can do whatever they want. If you really want to do something and are willing to put in the hard work it takes, then someday–bingo, you've done it!"

SCIENCE GAZETTE

ROBOTS:

Do They Signal Automation or Unemployment?

Sparks fly as a worker welds parts to an automobile body. Farther down the assembly line, another worker trims and grinds the weld joints. Beyond that, still another worker sprays paint on the car.

Up and down the assembly line, not a word is spoken as the workers perform their tasks. The workers do not pause, yawn, blink, or look at each other. "That's just not human," you might say. And you would be right. For these workers are robots!

Robots are becoming more and more common in assembly-line jobs. Dr. Harley Shaiken of the Massachusetts Institute of Technology has predicted that 32,000 robots may one day replace 100,000 automobile-industry workers. Is this prediction likely to come true?

ROBOT REVOLUTION

Already, thousands of robots are used in factories all over the United States. Every day more are being put to work. Several fully automated factories are now being tested. In such factories, all of the production and assembly is done by machines. One result of this "robot revolution" is increased unemployment in industrial regions. Workers in these areas are demanding that industry leaders slow down the switch to robots.

But, as many company executives point out, robots often perform jobs that are boring and repetitive, as well as jobs that may be hazardous to humans. For example, a robot may paint many thousands of cars and not be affected by inhaling paint fumes that may be dangerous for a human worker to inhale.

Other company executives point out that they must either use robots or lose business to companies that do. As Thomas B. Gunn of the Arthur D. Little Company puts it, "are you going to reduce your work force by 25 percent by putting in robots, or by 100 percent by going out of business?" And James Baker of the General Electric Corporation puts it this way: "U.S. business has three choices in the 90s...automate, emigrate (leave), or evaporate (go out of business)."

NEW JOBS OR FEW JOBS?

Some industry experts believe that robots and computer systems will create many new jobs. "In the past," George Brosseau of the National Science Foundation explains, "whenever a new technology has been introduced, it has always generated more jobs than it has displaced. But we don't know whether that's true of robot technology. There's no question but that new jobs will be created, but will there be enough to offset the loss?" he adds.

James S. Albut of the U.S. National Institute of Standards and Technology says yes. "Robots can improve productivity and create many new jobs," he has written.

Others reply that although using robots may create more jobs in the long run, they cause job losses first. These people stress that it is up to government, business, and labor to teach people new skills and to find new jobs for workers who have been replaced by robots. Such efforts will ease some of the strain that accompanies the implementation of new and valuable technology.

Do you think United States companies should rapidly move ahead with the development and use of robots even at the risk of some unemployment and worker hardship? Or should the switch from human power to robot power be done slowly in order to lessen the impact on workers? Before you answer, carefully consider the effects of your decision.

HYPERSONIC PLANES:

Flying Faster Than the Speed of Sound

"**H**urry, Sandy, it's time to leave," Mrs. Wilson said into the computer-roomspeaker.

Sandy heard her mother's voice jump out of the speaker in her bedroom. Sandy was late again. She and her mother were on their way to the airport to pick up Sandy's sister, Maria. She had been visiting friends in New York City. But Sandy couldn't find her blue jacket.

Sandy pressed the button on her wristband. A computer voice said, "The time is now 12:15 PM." Maria's plane is leaving New York right now, Sandy thought. She'll be here in San Francisco in less than 45 minutes.

Grabbing a red sweater and racing out the door, Sandy joined her mother, who was waiting patiently in the family turbocar. As Sandy jumped in, her mom pressed the control stick. The car glided off the ground and toward the airport.

"You should have been born 100 years ago, in the 1980s," Mrs. Wilson said, glancing at her daughter with a smile. "Everything was much slower then."

Sandy gave her mother a puzzled look.

"What do you mean?" she asked.

"Today, a trip from New York to San Francisco on a hypersonic plane, better known as an HST, takes about one-half hour. That doesn't give you much time to get your wardrobe together," Mrs. Wilson teased. "But just 100 years ago, you couldn't fly from New York to San Francisco in less than five hours. In fact, just 160 years ago, you wouldn't have been able to fly from New York to San Francisco at all! You would have

had to travel by old-style trains that rolled on metal tracks or by cars that moved on rubber wheels along the ground."

"How long did that take?" Sandy asked.

"Oh, days," Mrs. Wilson answered. "Now HSTs can travel at a top speed of Mach 25. That's 25 times the speed of sound, or about 27,200 kilometers per hour. One hundred years ago, top speeds were less than 1000 kilometers per hour."

"No wonder trips took so long," Sandy said.

Sandy thought for a moment and then did some quick arithmetic in her head. "That means that Maria's flight took...five hours back then. Five hours!"

"There were also supersonic planes, or SSTs," said Mrs. Wilson, as she shifted the turbocar to a slower speed. Traffic was heavy. "SSTs were the step before the HSTs we have today. SSTs flew about twice the speed of sound, or about 2480 kilometers per hours. An SST, like the one they used to call the *Concorde*, could cross the Atlantic Ocean in four hours. That's much faster than a regular jet back then. But it's still slower than the one hour it takes to cross the Atlantic today in an HST."

"How about travelling across the United States?" Sandy asked.

"SSTs didn't, Mrs. Wilson explained. "When a plane nears the speed of sound, air bunches up in front of it. Pressure waves are created. When the plane passes the speed of sound, it breaks through these waves. Peo-

▲ This illustration shows an artist's conception of a hypersonic plane of the future.

ple on the ground hear a loud noise called a sonic boom. In the 1970s, when SSTs started to fly, people objected to the noise. SST flights were also bothered by environmental and economic concerns. So SST flights were prohibited over the United States. They were permitted only over oceans. We don't have those problems today. The streamlined design of the HST lessens the pressure waves. The HST also flies higher. So noise isn't as great."

"When did we start using HSTs?" Sandy asked.

"I can't remember the exact date," her mother said, shifting the turbocar to stop in the traffic. "They started developing the HST in the 1980s. They had small models of the plane, and they had tested new engines that would be needed for hypersonic speeds. Hypersonic speed is any speed greater than five times the speed of sound. But it took many years before the first plane was ready. That was around the year 2000."

"Everyone must have been extremely thrilled when the HST first took to the air," Sandy said.

"Well, yes and no. Some people were very pleased," her mom answered. "Lots of business was opening with countries in Asia. And the HST was seen as a way to lessen travel time. A flight from

Washington, D.C., to Tokyo originally might have taken 15 hours. Today on an HST, it takes only one. But some people were skeptical."

"Well, I can imagine that," Sandy laughed. "Flying 27,200 kilometers per hour is a lot different from flying 2480 kilometers per hour."

"True, but there were other differences," Sandy's mom said, leaning back to wait for the traffic to ease. "Our HST is no bigger than some of the jets that flew back then. It carries 500 passengers, the same as the old 747s, or jumbo jets. But when it takes off, it picks up speed faster and climbs higher. On most flights, it's actually in low orbit around the earth–at the edge of space. No passenger plane of the 1980s did that."

"You sure know a lot about the history of airplanes," Sandy said, smiling at her mom. "I guess it helps that you're an HST pilot."

"It helps with the history of flight," Sandy's mother answered. "But it doesn't help get us out of this traffic. Imagine, we can fly across the country in a half hour, but we can't make it across town to meet the plane on time."

"That's one thing from the old days that hasn't changed very much," Sandy said with a laugh.

San Francisco
45 minutes
New York
3 days
5+ hours

▲ This illustration compares the length of time required to travel from New York to San Francisco on a train, a conventional airplane, and the HST.

For Further Reading

If you have been intrigued by the concepts examined in this textbook, you may also be interested in the ways fellow thinkers—novelists, poets, essayists, as well as scientists—have imaginatively explored the same ideas.

Chapter 1: What Is Motion?

Cormier, Robert. *I Am the Cheese.* New York: Pantheon.

Lee, Harper. *To Kill a Mockingbird.* New York: Harper.

Sillitoe, Alan. *The Loneliness of the Long-Distance Runner.* New York: Knopf.

Chapter 2: The Nature of Forces

Stewart, Michael. *Monkey Shines.* New York: Freundlich.

Store, Josephine. *Green Is for Galaxy.* New York: Argo Books.

Wells, H.G. *The Invisible Man.* New York: Watermill Press.

Chapter 3: Forces in Fluids

Ferber, Edna. *Giant.* Garden City, NY: Doubleday.

Gallico, Paul. *The Poseidon Adventure.* New York: Coward.

Hemingway, Ernest. *The Old Man and the Sea.* New York: Scribner.

Chapter 4: Work, Power, and Simple Machines

Crichton, Michael. *The Great Train Robbery.* New York: Knopf.

Foster, Genevieve. *The Year of the Flying Machine: 1903.* New York: Scribner.

Gardner, Robert. *This Is the Way It Works: A Collection of Machines.* New York: Doubleday.

Chapter 5: Energy: Forms and Changes

Bograd, Larry. *Los Alamos Light.* New York: Farrar.

Bond, Nancy. *The Voyage Begun.* New York: Argo Books.

McDonald, Lucile. Windmills: *An Old-New Energy Source.* New York: Elsevier/Nelson Books.

Activity Bank

Welcome to the Activity Bank! This is an exciting and enjoyable part of your science textbook. By using the Activity Bank you will have the chance to make a variety of interesting and different observations about science. The best thing about the Activity Bank is that you and your classmates will become the detectives, and as with any investigation you will have to sort through information to find the truth. There will be many twists and turns along the way, some surprises and disappointments too. So always remember to keep an open mind, ask lots of questions, and have fun learning about science.

Chapter 1	**WHAT IS MOTION?**	
	FLYING HIGH	**138**
Chapter 2	**THE NATURE OF FORCES**	
	SMOOTH SAILING	**140**
	AT THE CENTER OF THE GRAVITY MATTER	**141**
	LIGHT ROCK	**142**
	PUTTING GRAVITY TO WORK	**143**
Chapter 3	**FORCES IN FLUIDS**	
	WATERING YOUR GARDEN GREEN	**145**
	DENSITY DAZZLERS	**147**
	BAFFLING WITH BERNOULLI	**149**
Chapter 5	**ENERGY: FORMS AND CHANGES**	
	CRAZY EIGHTS	**151**

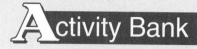

FLYING HIGH

Did you get into trouble the last time you launched a paper airplane across a room? Under the right circumstances (such as during a classroom activity), however, paper airplanes can be used to learn a lesson or two about motion. In this activity you will build and fly paper airplanes so that you can practice calculating speed and distance as well as learn a bit about designing an airplane.

What You Need

sheets of paper of various weights and sizes

stopwatch with a second hand

meterstick

string

adhesive tape

paper clips

stapler with staples

What You Do

1. As a group, construct a paper airplane using the materials provided. Design it any way that you would like.

2. Once all of the groups in your class have finished building their airplanes, get together to set up a flight-test area for the airplanes. This is the location in which you will hold an airplane-flying contest. It should be a large open space with few obstacles. You should also devise a set of rules by which the flights will be judged. For example, all flights should begin from the same spot. You may wish to place a piece of tape on the floor to mark the starting point for the flights. You should also decide when the time measurement begins (such as at the release of the airplane), and to where distance is measured (some airplanes will slide after they reach the floor). **CAUTION:** *The test area should be carefully controlled so that no one walks into the flight pattern. A pointed airplane nose can be quite dangerous.*

3. Now assign each group member one of the following roles: Thrower, Timer, Distance measurer, Recorder. Throw several test flights with your airplane. Each time measure the duration of the flight and the distance the airplane flew. Record this information in a data table similar to the one shown. Then use the information to calculate the airplane's speed. Take turns assuming the role of the thrower so that each group member gets a chance.

4. Now you are going to have a friendly competition with your classmates. When everyone is ready, take turns throwing your airplanes. Have independent recorders take the measurements and record them in another data table.

What You Saw

DATA TABLE

	Time (sec)	Distance (m)	Speed (m/sec)
Trial 1			
Trial 2			
Trial 3			

What You Learned

1. How did you calculate speed? What was the slowest speed for your group's airplane? The fastest? What about in the entire class?

2. How would you want your data to change if you are trying to decrease speed? (*Hint:* How would you change distance and time?)

3. Were the speeds you calculated actual or average speeds?

4. Did you notice anything about the slowest airplanes and the shape of their flight paths? You may need to see them flown again.

5. Compare the designs of the fastest airplanes with those of the slowest. How are they alike? Different? How would you redesign your airplane to make it move more quickly? More slowly?

The Next Step

Redesign your airplane to make it move faster. Repeat the activity, but this time see if the change in design achieves a faster speed.

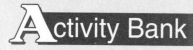

SMOOTH SAILING

If you have ever watched a rocket being launched, you know what a spectacular sight it can be. But what you may not have realized is that a rocket's blastoff can be described by Newton's third law of motion. Newton explained that for every action there is an equal and opposite reaction. In the case of a rocket, the burning fuel pushes out in one direction, forcing the rocket to move in the other direction. In this activity you will experiment with Newton's third law of motion by designing a "rocket-powered" boat.

Materials

1.89 L (half-gallon) milk or juice carton
scissors
balloon (hot-dog shaped)
bathtub or sink filled with water

Procedure

1. Wash the carton thoroughly. Using scissors, carefully cut one long side from the carton. You may find it helpful to begin your cut by poking a small hole in the carton with the sharp end of the scissors.

2. Again using the scissors, carefully cut a small hole near the center of the bottom of the carton. Do not make the hole too big.

3. Place the hot-dog shaped balloon in the carton and run its neck through the hole in the bottom of the carton.

4. Blow up the balloon but do not tie it. Instead, use your fingers to keep the neck of the balloon closed and the air in the balloon.

5. Place your balloon boat in a bathtub or sink filled with water. Let go of the balloon's neck and observe what happens.

Analysis and Conclusions

1. What happened when you released the balloon?

2. What did the boat use as fuel?

3. What would happen if you blew less air into the balloon? More air? Try it.

4. How is the balloon boat an example of Newton's third law of motion?

The Next Step

Predict what will happen to the boat's speed if you place some cargo in the boat. To find out, use small masses such as marbles, stones, or coins. Repeat the boat run several times, each time gradually increasing the mass of the cargo. Make sure that you blow the same amount of air into the balloon each time.

- What happens to the boat's speed and the distance it travels as you continue to add cargo to it?

- Which of Newton's laws of motion explains these observations?

AT THE CENTER OF THE GRAVITY MATTER

Have you ever balanced a ruler on your finger? If so, you know that it balances only when its center is placed on your finger. Placing any part of the ruler other than its center on your finger will not provide the desired result. This is because the center of the ruler is where the ruler's center of gravity is located. All objects have a center of gravity. The center of gravity of an object is the point on which gravity seems to pull. In reality, gravity pulls downward on every point on an object. Yet when the forces on all of the points are added together, it is as if the total force of gravity pulls on only one point—the center of gravity.

Not all objects are quite as predictable as a ruler. In fact, the center of gravity for some objects is not even on the object! Do you know where your center of gravity is? Find out by gathering a bandana or handkerchief and a wooden block or chalkboard eraser and performing the following steps.

What to Do

1. Stand with the entire left side of your body against a wall. Make sure your left foot is up against the wall. Now try to lift your right foot. What happens?

2. Now stand with your back against a wall. Be sure your heels are touching the wall. Drop a bandana or handkerchief just in front of your toes. Try to pick it up without bending your knees or moving your feet. Describe what happens.

3. Place a mat or blanket on the floor. Get down on your elbows and knees on top of the mat or blanket. Place your elbows on the floor right in front of your knees. At the tip of your middle fingers, place a wooden block or blackboard eraser on its edge. With your hands

behind your back, lean forward and try to knock the block over with your nose. Can you do it?

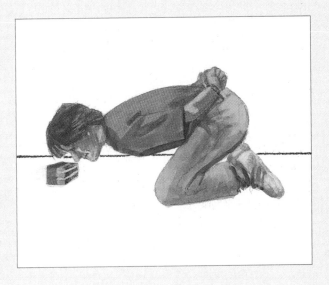

What to Think About

Make a chart showing the results for your class—who was able to hit the eraser and who wasn't. Overall, are girls or boys more successful? Why do you think this is so?

Try Again

Repeat step 3 but this time place weights in your back pockets or on your ankles. If weights are not available, have a friend hold your ankles down. Does this change your results?

What to Understand and Apply

1. Why do you think it is important for the center of gravity of a car or truck to be located in a proper position? What can happen if it is too high?

2. Why do you think tightrope walkers use long poles to help balance themselves?

LIGHT ROCK

In the Activity Bank activity on page 141, you learned about the center of gravity—what it is and how it affects you in many different ways. You can alter the center of gravity of an object only by altering the object. In this activity you will investigate the rather interesting results of altering an object's center of gravity.

Materials You Will Need

standard cork from a bottle (2.5 cm to 3 cm in diameter)
sewing needle
knitting needle or bamboo skewer
2 candles about 8 cm long and 1.5 cm in diameter
2 drinking glasses of equal height
several sheets of newspaper
box of matches

Procedure

1. Push the sewing needle sideways through the cork. **Note:** *Be careful not to break the needle and do not place your hand in such a way that the needle can poke you as it goes through the cork.*

2. Carefully slide the bottom of one candle onto one end of the sewing needle so that the needle extends to the center of the candle and the candle extends out

from the cork. See the accompanying diagram. Repeat this procedure for the other candle.

3. Push the knitting needle lengthwise (bottom to top) through the cork. Be careful not to hit the sewing needle. (You may need to direct the knitting needle slightly above or beneath the sewing needle.) Place the setup over two drinking glasses that have been turned upside down. Make sure the glasses are placed on top of several sheets of newspaper.

4. Light both candles. **CAUTION:** *Be careful when using matches.* Observe the setup for several minutes.

Observations and Conclusions

1. What happens when the candles are lighted? Why?

2. How could you alter the center of gravity of other objects, such as cars?

PUTTING GRAVITY TO WORK

If someone asked you what gravity does, you might say that it is responsible for pulling you to the ground when you stumble over a stone or fall off your bicycle. While these answers are painfully correct, it is important to realize that gravity is more often an extremely helpful force. In addition to keeping you from flying off the Earth's surface, gravity has important applications in the operation of many modern devices. In this activity you will find out how gravity can be used to test the strengths of various materials.

Materials

ring stand and ring
masking tape
plastic or paper cup
strips of several different kinds of paper (14 cm x 3 cm); examples include paper towels, tissue paper, toilet paper, wrapping paper, writing paper, and typing paper
250-mL beaker
sand, 250 mL
balance
several sheets of newspaper

Procedure 🜊

1. Tape one end of the first paper strip to the ring. Tape the other end to the rim

of the plastic or paper cup. Spread some newspapers below the cup.

2. Fill the beaker with sand. Slowly pour the sand from the beaker into the cup. Stop pouring when the paper begins to tear. Be careful not to let the cup fall because the sand will spill out of it.

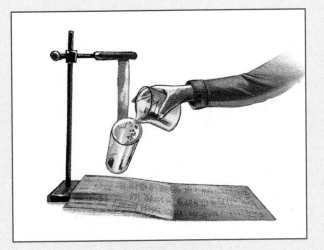

3. Remove the cup with the sand in it and discard the paper strip. Use the balance to measure the mass of the cup with the sand in it. On a separate sheet of paper, make a data table similar to the one shown on the next page. Make sure your data table includes all of the different types of paper you have. Record the mass in the appropriate column in your data table and return the sand to its container.

4. Repeat steps 1, 2, and 3 using each of the different strips of paper. Make sure the cup is clean each time.

5. Calculate the weight held by each paper strip by multiplying the mass you recorded by 9.80 N/kg. (**Note:** *Make sure you convert the masses from grams to kilograms first by dividing them by 1000.*) *(continued)*

Observations and Calculations

DATA TABLE

Sample	Mass (g)	Weight (N)
A		
B		
C		
D		

Analysis

1. Which type of paper supports the greatest weight? The least?
2. How did the investigation differentiate the types of paper by their strengths?
3. Why is gravity a good choice of force to use to measure the strength of the different materials?

The Next Step

Repeat the experiment but this time cut the papers at a 90° angle to the direction in which you cut them the first time. For example, if you cut a letter size sheet of paper lengthwise, cut it widthwise this time. How does this affect your results? Can you propose a hypothesis as to why?

WATERING YOUR GARDEN GREEN

Have you ever grown a plant or kept a garden? Whether you have or have not, the first item you probably think is required for such an endeavor is soil. After all, how can you grow a plant without soil? But you can! In this activity you will grow a small garden without soil—thanks to the principles of fluid pressure.

Materials

large plastic container	small plastic container with a lid
nail	hammer
nylon stocking	twist tie
white glue	Perlite or sand
water	houseplant fertilizer
spoon	seedlings
plastic aquarium tubing (60 cm)	

Procedure

1. Using the hammer and nail, punch a hole in one side of each container near the bottom. The hole must be large enough to accommodate the plastic tubing. Take care, however, not to crack the containers. If you find it easier, you may want to use clean milk or juice cartons instead of plastic containers. Make another hole in the center of the lid of the smaller container.

2. Place a piece of nylon stocking over one end of the aquarium tubing. Secure it there with the twist tie. Push this end into the hole in the large container. Push the other end of the aquarium tubing into the bottom hole in the small container.

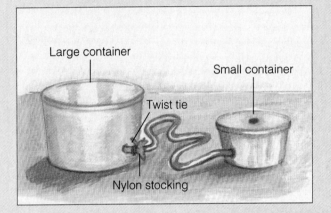

Large container

Small container

Twist tie

Nylon stocking

3. Surround the holes with glue so that they become watertight. Let the glue dry thoroughly.

4. Fill the large container with wet Perlite and the small container with water. Add a pinch of fertilizer to the water in the small container, stir, and put on the lid. Place the seedlings in the large container. You have completed the construction of your garden.

5. To make your garden grow, put it on a sunny windowsill. Feed it once a day by lifting the water container higher than the plant container. When the water container empties, place it lower than the plant container and it will fill up again. There, you have fed your garden for the day.

6. Once a month add fresh water to the water container and stir in a pinch of fertilizer. Enjoy the results of your "green thumb"!

(continued)

Analysis

1. Why does lifting the small container affect the plants in the large container?

2. Why do you need to put fertilizer in the water?

3. What would happen if a small leak or hole developed in the tubing?

4. What is the purpose of using the nylon stocking?

5. Can you think of a common device that operates in a manner that is somewhat similar to this water garden?

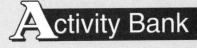

DENSITY DAZZLERS

One scorching summer afternoon, while out in a row boat with your friend Alex and your dog Sam, you learn a quick lesson about density. It all begins when Sam jumps up to bark at a duck, then Alex stands up to grab Sam, and before you know it, you are all in the water. After putting on your life vests, you begin to look around for the rest of your belongings. The boat is floating nearby—upside down, unfortunately. The soccer ball you brought is floating beside the plastic utensils that were with your lunch. But where is your lunch? And where is your radio? Why did they sink if everything, and everyone, else is floating?

The answer to why some objects float and others do not has to do with density. An object can float only if it is less dense than the substance it is in. In this activity you will complete your own investigations into density. But don't worry—you won't get wet!

Materials

2 250-mL beakers
cooking oil (about 125 mL)
ice cube
salt
spoon or small sheet of tissue paper (optional)
hard-boiled egg or raw potato
medicine dropper
dishwashing liquid
food coloring

Procedure 🔬

1. Fill a beaker half-full with cooking oil. Very gently place an ice cube on the surface of the oil. What happens to the ice cube? Watch the ice cube for the next 15 to 20 minutes. What happens as the ice cube melts?

2. Use the dishwashing liquid to thoroughly clean the beaker. Fill the beaker half-full with water. Make sure you know the volume of water you put in.

3. Dissolve plenty of salt in the water. The amount of salt will vary depending on exactly how much water you use. Stop adding salt when the water becomes cloudy.

4. Add the same amount of water you used in step 2 to another beaker. Do not add salt this time. Slowly pour the water into the beaker containing the salt water in such a way that it does not mix with the salt water. You may need to pour it over a spoon or sheet of tissue paper so that it hits the salt water more gently.

5. Gently place an egg or small potato in the beaker. Describe and draw what you see.

6. Clean out the two beakers. Add a small amount of hot tap water, about 10 mL, to one of the beakers. Add food coloring to the hot water. You can choose any color that you wish, but make sure you add enough food coloring to the water so that you can see the color well.

(continued)

7. Fill the other beaker with cold tap water.

8. Use a medicine dropper to pick up a few drops of the hot colored water.

9. Place the tip of the medicine dropper in the middle of the cold water. Now slowly squeeze a drop of the hot colored water into the cold water. Describe and draw a picture of what you see.

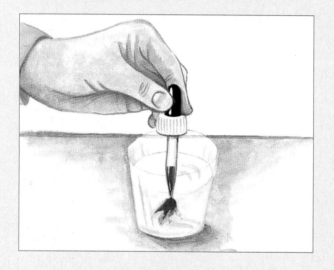

Analysis and Conclusions

1. Explain the observations you made when watching the ice cube you placed in the oil in step 1.

2. Explain your observations regarding the egg in the beaker of water.

3. What does your experiment tell you about the density of hot water as compared to that of cold water?

4. Predict what will happen if you repeat steps 6 to 9 but this time add cold colored water to hot water. Try it. Are you correct?

BAFFLING WITH BERNOULLI

Have you ever tried to perform a magic trick for your friends or family? What makes a magic trick work is knowing the secret. Well, now you can use Bernoulli's principle to amaze and baffle others who don't know about forces and fluids, as you do!

What You Will Need

2 drinking straws
drinking glass
2 balloons
fine thread
tape
small funnel

Ping-Pong ball
thumbtack
halved index card
wooden spool (the kind used for cotton thread)

What You Will Do

Part A

1. For the first "trick," hold a drinking straw upright in a glass of water so that the bottom of the straw is slightly above the bottom of the glass.

2. Use a second straw to blow a stream of air across the top of the first straw. (If you have only one straw, you can cut it in half to make two straws.) Vary the force with which you blow the stream of air.

■ What happens when you blow into the second straw?

■ What happens when you blow harder?

■ Can you explain why this "trick" is possible?

Part B

If your audience was impressed with what happened with the straws, they'll really be excited by this next "trick."

1. Blow up two balloons. Tie about 60 cm of thread to each one. Use tape to hang them from the top of a door

frame, light fixture, or low ceiling, about 5 cm apart.

2. Have your audience predict what will happen to the balloons if you blow a stream of air between them. Go ahead and do it. (You may find it easier to blow through a straw that you hold between the balloons.)

■ What happens? Was your audience correct in their prediction?

Part C

In this next display, you will set out to defy gravity. Do you think it can be done?

1. Hold the small funnel upright and place a Ping-Pong ball in it. Blow through the narrow end of the funnel.

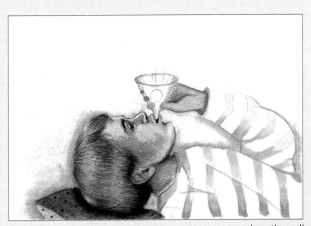

(continued)

- Can you blow the ball out? Why or why not?

2. Now turn the funnel downward and hold the Ping-Pong ball inside it with your hand. Blow on the funnel again and let go of the ball. What do you think will happen when you let go?

- What does happen? Why?

Part D

This last "trick" will surprise everyone—maybe even you!

1. Push a thumbtack through the middle of a halved index card.

2. Hold the index card and tack under the wooden spool so that the pin projects into the hole of the spool. Blow hard down through the other hole in the spool and let go of the card. What do you expect will happen?

- What happens to the index card?

The Next Step

Design a poster that relates each of these "tricks" to Bernoulli's principle. Together with your classmates, come up with additional examples of Bernoulli's "trickery" and add them to your poster.

CRAZY EIGHTS

Have you ever watched the pendulum swinging in a grandfather clock? There is something about it that truly captures one's attention. The pendulum always swings as far to the right as it does to the left in a regular repeating pattern. But what do you think would happen if a pendulum was made to swing on another pendulum? In this activity you will find out.

Materials

plastic funnel
heavy wire (about 120 cm long)
sheet of paper
several sheets of newspaper or wrapping paper
sand (or salt or sugar)
3 lengths of string, about 1 m each
scissors
metric ruler
compass
adhesive tape

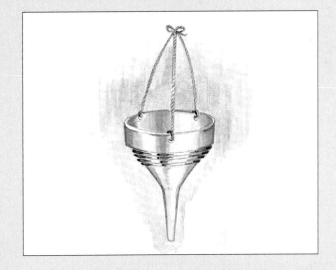

Procedure

1. Use the scissors to poke three small holes in the funnel near the edge of the larger end. The holes should be evenly spaced around the opening.

2. Tie an end of one string to one of the holes. Repeat this for the remaining two strings and two holes.

3. Wrap the heavy wire around the rim of the funnel to make it heavy. You may wish to have a partner hold the strings out of the way so that they do not become tangled in the wire.

4. Use the compass to draw a circle on the sheet of paper. The radius of the circle (the distance from the center to the edge) should be slightly less than the height of your funnel. Cut out the circle. Then make one straight cut from the edge to the center.

5. Fold the circular piece of paper into a cone that will fit into your funnel. When you have found the right size, tape it so that it holds it shape. Then cut a small hole in it at the point and place the cone in the funnel.

6. Arrange two chairs or desks about 60 cm apart with several sheets of newspaper or wrapping paper spread on the floor between them. Tie one string to the back of each chair or desk. With your fingers, pull these two strings together about 30 cm above the funnel. Pull the remaining string up to this point. Secure the three strings together at this point with a narrow piece of adhesive tape. (If the adhesive tape is not strong enough, you may want to use a small piece of wire or twist tie. Do not tie the strings together because you will need to adjust this length later on.)

(continued)

7. Cover the narrow funnel opening with your finger. Pour fine sand (or salt or sugar) into the funnel. Pull the funnel to the side and let go, removing your finger as you do so.

- What do you see happening to the sand or salt?

- What would happen if the funnel was supported by only two strings?

8. Make a chart listing the distance between the funnel and the gather where you placed the adhesive tape. Next to that information, draw the pattern created by the sand (or salt or sugar). Leave room on your chart for several more trials.

9. Carefully pick up the edges of the newspaper or wrapping paper so that the sand moves to the center. Then pour it back into its container. Repeat the activity several times, each time changing the length of the string below the gather where you placed the adhesive tape by moving the tape up or down. Add the data and the drawings to your chart.

- When you are finished, share your observations with those of your classmates. How are your observations similar to theirs? How are they different? Did anyone find any surprising or unusual observations?

- Do you think your results would change if you pulled the funnel back further to start the motion? What about if you pulled it less? Why? Try it and see.

Appendix A

The metric system of measurement is used by scientists throughout the world. It is based on units of ten. Each unit is ten times larger or ten times smaller than the next unit. The most commonly used units of the metric system are given below. After you have finished reading about the metric system, try to put it to use. How tall are you in metrics? What is your mass? What is your normal body temperature in degrees Celsius?

Commonly Used Metric Units

Length   The distance from one point to another

meter (m)	A meter is slightly longer than a yard.
	1 meter = 1000 millimeters (mm)
	1 meter = 100 centimeters (cm)
	1000 meters = 1 kilometer (km)

Volume   The amount of space an object takes up

liter (L)	A liter is slightly more than a quart.
	1 liter = 1000 milliliters (mL)

Mass   The amount of matter in an object

gram (g)	A gram has a mass equal to about one paper clip.
	1000 grams = 1 kilogram (kg)

Temperature   The measure of hotness or coldness

degrees	0°C = freezing point of water
Celsius (°C)	100°C = boiling point of water

Metric–English Equivalents

2.54 centimeters (cm) = 1 inch (in.)
1 meter (m) = 39.37 inches (in.)
1 kilometer (km) = 0.62 miles (mi)
1 liter (L) = 1.06 quarts (qt)
250 milliliters (mL) = 1 cup (c)
1 kilogram (kg) = 2.2 pounds (lb)
28.3 grams (g) = 1 ounce (oz)
$°C = 5/9 \times (°F - 32)$

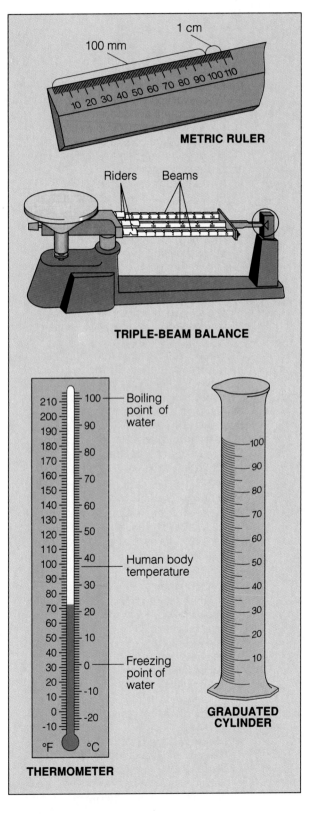

METRIC RULER

TRIPLE-BEAM BALANCE

THERMOMETER

GRADUATED CYLINDER

Glassware Safety

1. Whenever you see this symbol, you will know that you are working with glassware that can easily be broken. Take particular care to handle such glassware safely. And never use broken or chipped glassware.
2. Never heat glassware that is not thoroughly dry. Never pick up any glassware unless you are sure it is not hot. If it is hot, use heat-resistant gloves.
3. Always clean glassware thoroughly before putting it away.

Fire Safety

1. Whenever you see this symbol, you will know that you are working with fire. Never use any source of fire without wearing safety goggles.
2. Never heat anything—particularly chemicals—unless instructed to do so.
3. Never heat anything in a closed container.
4. Never reach across a flame.
5. Always use a clamp, tongs, or heat-resistant gloves to handle hot objects.
6. Always maintain a clean work area, particularly when using a flame.

Heat Safety

Whenever you see this symbol, you will know that you should put on heat-resistant gloves to avoid burning your hands.

Chemical Safety

1. Whenever you see this symbol, you will know that you are working with chemicals that could be hazardous.
2. Never smell any chemical directly from its container. Always use your hand to waft some of the odors from the top of the container toward your nose—and only when instructed to do so.
3. Never mix chemicals unless instructed to do so.
4. Never touch or taste any chemical unless instructed to do so.
5. Keep all lids closed when chemicals are not in use. Dispose of all chemicals as instructed by your teacher.

6. Immediately rinse with water any chemicals, particularly acids, that get on your skin and clothes. Then notify your teacher.

Eye and Face Safety

1. Whenever you see this symbol, you will know that you are performing an experiment in which you must take precautions to protect your eyes and face by wearing safety goggles.
2. When you are heating a test tube or bottle, always point it away from you and others. Chemicals can splash or boil out of a heated test tube.

Sharp Instrument Safety

1. Whenever you see this symbol, you will know that you are working with a sharp instrument.
2. Always use single-edged razors; double-edged razors are too dangerous.
3. Handle any sharp instrument with extreme care. Never cut any material toward you; always cut away from you.
4. Immediately notify your teacher if your skin is cut.

Electrical Safety

1. Whenever you see this symbol, you will know that you are using electricity in the laboratory.
2. Never use long extension cords to plug in any electrical device. Do not plug too many appliances into one socket or you may overload the socket and cause a fire.
3. Never touch an electrical appliance or outlet with wet hands.

Animal Safety

1. Whenever you see this symbol, you will know that you are working with live animals.
2. Do not cause pain, discomfort, or injury to an animal.
3. Follow your teacher's directions when handling animals. Wash your hands thoroughly after handling animals or their cages.

$\mathbf{A}$ ppendix C

One of the first things a scientist learns is that working in the laboratory can be an exciting experience. But the laboratory can also be quite dangerous if proper safety rules are not followed at all times. To prepare yourself for a safe year in the laboratory, read over the following safety rules. Then read them a second time. Make sure you understand each rule. If you do not, ask your teacher to explain any rules you are unsure of.

Dress Code

1. Many materials in the laboratory can cause eye injury. To protect yourself from possible injury, wear safety goggles whenever you are working with chemicals, burners, or any substance that might get into your eyes. Never wear contact lenses in the laboratory.

2. Wear a laboratory apron or coat whenever you are working with chemicals or heated substances.

3. Tie back long hair to keep it away from any chemicals, burners and candles, or other laboratory equipment.

4. Remove or tie back any article of clothing or jewelry that can hang down and touch chemicals and flames.

General Safety Rules

5. Read all directions for an experiment several times. Follow the directions exactly as they are written. If you are in doubt about any part of the experiment, ask your teacher for assistance.

6. Never perform activities that are not authorized by your teacher. Obtain permission before "experimenting" on your own.

7. Never handle any equipment unless you have specific permission.

8. Take extreme care not to spill any material in the laboratory. If a spill occurs, immediately ask

your teacher about the proper cleanup procedure. Never simply pour chemicals or other substances into the sink or trash container.

9. Never eat in the laboratory.

10. Wash your hands before and after each experiment.

First Aid

11. Immediately report all accidents, no matter how minor, to your teacher.

12. Learn what to do in case of specific accidents, such as getting acid in your eyes or on your skin. (Rinse acids from your body with lots of water.)

13. Become aware of the location of the first-aid kit. But your teacher should administer any required first aid due to injury. Or your teacher may send you to the school nurse or call a physician.

14. Know where and how to report an accident or fire. Find out the location of the fire extinguisher, phone, and fire alarm. Keep a list of important phone numbers—such as the fire department and the school nurse—near the phone. Immediately report any fires to your teacher.

Heating and Fire Safety

15. Again, never use a heat source, such as a candle or burner, without wearing safety goggles.

16. Never heat a chemical you are not instructed to heat. A chemical that is harmless when cool may be dangerous when heated.

17. Maintain a clean work area and keep all materials away from flames.

18. Never reach across a flame.

19. Make sure you know how to light a Bunsen burner. (Your teacher will demonstrate the proper procedure for lighting a burner.) If the flame leaps out of a burner toward you, immediately turn off the gas. Do not touch the burner. It may be hot. And never leave a lighted burner unattended!

20. When heating a test tube or bottle, always point it away from you and others. Chemicals can splash or boil out of a heated test tube.

21. Never heat a liquid in a closed container. The expanding gases produced may blow the container apart, injuring you or others.

22. Before picking up a container that has been heated, first hold the back of your hand near it. If you can feel the heat on the back of your hand, the container may be too hot to handle. Use a clamp or tongs when handling hot containers.

Using Chemicals Safely

23. Never mix chemicals for the "fun of it." You might produce a dangerous, possibly explosive substance.

24. Never touch, taste, or smell a chemical unless you are instructed by your teacher to do so. Many chemicals are poisonous. If you are instructed to note the fumes in an experiment, gently wave your hand over the opening of a container and direct the fumes toward your nose. Do not inhale the fumes directly from the container.

25. Use only those chemicals needed in the activity. Keep all lids closed when a chemical is not being used. Notify your teacher whenever chemicals are spilled.

26. Dispose of all chemicals as instructed by your teacher. To avoid contamination, never return chemicals to their original containers.

27. Be extra careful when working with acids or bases. Pour such chemicals over the sink, not over your workbench.

28. When diluting an acid, pour the acid into water. Never pour water into an acid.

29. Immediately rinse with water any acids that get on your skin or clothing. Then notify your teacher of any acid spill.

Using Glassware Safely

30. Never force glass tubing into a rubber stopper. A turning motion and lubricant will be helpful when inserting glass tubing into rubber stoppers or rubber tubing. Your teacher will demonstrate the proper way to insert glass tubing.

31. Never heat glassware that is not thoroughly dry. Use a wire screen to protect glassware from any flame.

32. Keep in mind that hot glassware will not appear hot. Never pick up glassware without first checking to see if it is hot. See #22.

33. If you are instructed to cut glass tubing, fire-polish the ends immediately to remove sharp edges.

34. Never use broken or chipped glassware. If glassware breaks, notify your teacher and dispose of the glassware in the proper trash container.

35. Never eat or drink from laboratory glassware. Thoroughly clean glassware before putting it away.

Using Sharp Instruments

36. Handle scalpels or razor blades with extreme care. Never cut material toward you; cut away from you.

37. Immediately notify your teacher if you cut your skin when working in the laboratory.

Animal Safety

38. No experiments that will cause pain, discomfort, or harm to mammals, birds, reptiles, fishes, and amphibians should be done in the classroom or at home.

39. Animals should be handled only if necessary. If an animal is excited or frightened, pregnant, feeding, or with its young, special handling is required.

40. Your teacher will instruct you as to how to handle each animal species that may be brought into the classroom.

41. Clean your hands thoroughly after handling animals or the cage containing animals.

End-of-Experiment Rules

42. After an experiment has been completed, clean up your work area and return all equipment to its proper place.

43. Wash your hands after every experiment.

44. Turn off all burners before leaving the laboratory. Check that the gas line leading to the burner is off as well.

Glossary

Pronunciation Key

When difficult names or terms first appear in the text, they are respelled to aid pronunciation. A syllable in SMALL CAPITAL LETTERS receives the most stress. The key below lists the letters used for respelling. It includes examples of words using each sound and shows how the words would be respelled.

Symbol	Example	Respelling
a	hat	(hat)
ay	pay, late	(pay), (layt)
ah	star, hot	(stahr), (haht)
ai	air, dare	(air), (dair)
aw	law, all	(law), (awl)
eh	met	(meht)
ee	bee, eat	(bee), (eet)
er	learn, sir, fur	(lern), (ser), (fer)
ih	fit	(fiht)
igh	mile, sigh	(mighl), (sigh)
oh	no	(noh)
oi	soil, boy	(soil), (boi)
oo	root, tule	(root), (rool)
or	born, door	(born), (dor)
ow	plow, out	(plow), (owt)

Symbol	Example	Respelling
u	put, book	(put), (buk)
uh	fun	(fuhn)
yoo	few, use	(fyoo), (yooz)
ch	chill, reach	(chihl), (reech)
g	go, dig	(goh), (dihg)
j	jet, gently, bridge	(jeht), (JEHNT-lee), (brihj)
k	kite, cup	(kight), (kuhp)
ks	mix	(mihks)
kw	quick	(kwihk)
ng	bring	(brihng)
s	say, cent	(say), (sehnt)
sh	she, crash	(shee), (krash)
th	three	(three)
y	yet, onion	(yeht), (UHN-yuhn)
z	zip, always	(zihp), (AWL-wayz)
zh	treasure	(TREH-zher)

acceleration: rate of change in velocity

Archimedes' principle: explanation that says that the buoyant force on an object is equal to the weight of the fluid displaced by the object

Bernoulli's principle: explanation that the pressure in a moving stream of fluid is less than the pressure in the surrounding fluid

buoyant (BOI-uhnt) **force:** upward force in a fluid that exists because the pressure of a fluid varies with depth

chemical energy: energy that bonds atoms or ions together

density: mass of a substance divided by its volume

efficiency: comparison of work input to work output

electromagnetic energy: energy associated with moving charges

energy: ability to do work

energy conversion: change of energy from one form to another

force: push or pull that gives energy to an object, sometimes causing a change in the motion of the object

frame of reference: background or point that is assumed to be stationary and is used for comparison when motion is described

friction: force that acts in the opposite direction of motion; will cause an object to slow down and finally stop

fulcrum: fixed pivot point of a lever

gravitational potential energy: potential energy that is dependent on height above the Earth's surface

gravity: force of attraction that depends on the mass of two objects and the distance between them; responsible for accelerating an object toward the Earth

heat energy: energy involved in the internal motion of particles of matter

hydraulic device: machine that takes advantage of the fact that pressure is transmitted equally in all directions in a liquid; obtains a large force on a large piston by applying a small force with a small piston

inclined plane: flat slanted surface that multiplies force

inertia (ihn-ER-shuh)**:** property of matter that tends to resist any change in motion

joule: unit of work and energy; 1 newton-meter

kinetic energy: energy of motion

Law of Conservation of Energy: law that states that energy is neither created nor destroyed by ordinary means

Law of Universal Gravitation: law that states that all objects in the universe attract each other by the force of gravity

lever: rigid bar free to move about a single point; may be first-class, second-class, or third-class depending on the positions of the effort force, resistance force, and fulcrum

machine: device that makes work easier by changing force and distance or by changing the direction of a force

mechanical advantage: number of times a machine multiplies the effort force

mechanical energy: energy associated with motion

momentum: mass of an object times its velocity; determines how difficult it is to stop the object's motion

motion: change in position in a certain amount of time

newton: unit of force; 1 kg × 1 m/sec/sec

nuclear energy: energy found in the nucleus of an atom

potential energy: energy of shape or position; stored energy

power: rate at which work is done

pressure: force that particles of a fluid exert over a certain area due to their weight and motion

pulley: rope, belt, or chain wrapped around a wheel; can either change the amount of force or the direction of the force

screw: inclined plane wrapped around a central bar to form a spiral

speed: rate at which an object moves

velocity: description of speed in a given direction

watt: unit of power; 1 joule per second

wedge: inclined plane that moves

wheel and axle: machine made up of two circular objects of different sizes; a force is applied to the wheel and transferred to the axle

work: force acting over a distance to move an object

work input: work that goes into a machine; effort force exerted over a distance

work output: work that comes out of a machine; output force exerted over a distance

Index

Acceleration, S21–24, S41
  finding acceleration of object, S22
  and gravity, S48–49
  negative acceleration, S22
  Newton's law about, S42, S43–44, S54
Air, buoyancy, S72
Air pressure, S63
  principle of unequal pressure and, S64–65
Air resistance, falling objects, S50–51
Archimedes' principle, S70–71
Average speed, S18–19

Balance, sense of and ear, S55
Bernoulli, Daniel, S74
Bernoulli's principle, S74–76, S77
Buoyancy, S70–73
  air, S72
  Archimedes' principle, S70–71
  at certain depth, S72–73
  and density, S71–72
  floating objects, S71–73
  nature of, S70–71

Chemical energy, S110, S114, S117
Circular motion, S24
Compound machines, S101
  definition of, S101
  examples of, S101
Conservation of energy, S119–121
  law of, S119–120
Conservation of momentum, S27–28
Constant speed, S16–18
Constant velocity, S41
Curve balls, S77

Deceleration, S22
Density, and buoyancy, S71–72

Efficiency, of machine, S91–92
Egyptians, S83
Einstein, Albert, S119–120
Electromagnetic energy, S110, S117, S118
Energy
  chemical energy, S110
  conservation of energy, S119–121
  definition of, S109
  electromagnetic energy, S110

heat energy, S109–110
kinetic energy, S112–113
mechanical energy, S109
nature of, S108–109, S124–125
nuclear energy, S110
potential energy, S113–115
study of physics and, S121–123
Energy conversions
  kinetic–potential energy conversions, S116
  types of, S117–118

Floating, buoyancy, S71–73
Fluid friction, S40
Fluid pressure, S62–69
  and depth of fluid, S68–69
  and gravity, S68–69
  principle of unequal pressure and, S64–65
Fluids. *See* Forces in fluids
Force, S121–122
  combining forces, S37–38
  definition of, S36
  friction, S38–40
  gravity, S47–54
  nature of force, S36
  Newton's laws of motion, S41–47, S54
  and work, S84–85
Forces in fluids
  Bernoulli's principle, S74–76
  buoyancy, S70–73
  fluid pressure, S62–65
  hydraulic devices, S65–67
Frames of reference, S12–13
Friction, S38–40, S41, S110
  definition of, S38
  fluid friction, S40
  nature of, S38–40
  rolling friction, S39–40
  sliding friction, S39, S40
Fulcrum, S95, S96

Galileo, S47–48
Gravitational potential energy, relationship to weight/height, S115
Gravity, S47–54
  air resistance, S50–51
  falling objects, S48–49
  and fluid pressure, S68–69
  Galileo's discovery, S47–48
  mass, S52, S54
  Newton's law of universal gravitation, S51–52
  weight, S53–54

Heat energy, S109–110
Height, and gravitational potential energy, S115
Hydraulic devices, S65–67
  examples of, S67
  principle of, S66–67

Inclined plane, S92–93
Inertia
  definition of, S41
  experience of, S41–42
  Newton's law, S42–43

Joules (J), S85, S109

Kepler, Johannes, S51
Kilowatts (kW), S87
Kinetic energy, S112–113
  definition of, S112
  kinetic–potential energy conversions, S116
  relationship to mass/velocity, S112–113

Laws of motion, S42–46
  concept of inertia, S42–43
  force/mass/acceleration relationship, S44, S54
  forces and opposite forces, S45–46
Law of universal gravitation, S51–52
Lever, S95–98

Machine
  definition of, S89
  ease of work and, S89–91
  efficiency of, S91–92
  mechanical advantage, S92
  See also Compound machines; Simple machines
Mass
  and gravity, S52, S54
  and kinetic energy, S112–113
  Newton's law about, S43–44, S54
Mechanical advantage, machine, S92
Mechanical energy, S109, S117, S118
Momentum, S25–28, S121
  conservation of momentum, S27–28
  definition of, S25
  finding momentum of object, S25–26

Motion
    ancient Greeks on, S41
    circular motion, S24
    definition of, S14
    frames of reference, S12–13
    laws of motion, S41–47
    momentum, S25–28
    speed, S14–19
    velocity, S19–24
Negative acceleration, S22
Newton, Isaac, S41, S48
Newton (N), S44, S53, S63
Newton-meter (N-m), S85
Newton's laws. *See* Laws of Motion;
    Law of universal gravitation
Nuclear energy, S110

Photography, S29
Physics, knowledge about energy
    and, S121–123
Potential energy, S113–115
    definition of, S113
    kinetic–potential energy
        conversions, S116
Power, S86–87, S122
    definition of, S86
    measurement of, S86

Pressure
    air pressure, S63, S64–65
    Bernoulli's principle, S74–76
    definition of, S62
    differences in, S63–65
    finding pressure, S62
    fluid pressure, S62–65
    measure of, S63
Pulley, S98–99

Rolling friction, S39–40

Screw, S94–95
Simple machines
    inclined plane, S92–93
    lever, S95–98
    pulley, S98–99
    screw, S94–95
    wedge, S93–94
    wheel and axle, S100–101
Sliding friction, S39, S40
Solar energy, S117
Speed, S14–19, S121
    average speed, S18–19
    constant speed, S16–18
    definition of, S15
    finding speed of objects, S15–16

Universal gravitation, law of,
    S51–52

Velocity
    acceleration, S21–24
    combining velocities, S19–20
    deceleration, S22
    definition of, S19
    and kinetic energy, S112–113
Vinci, Leonardo da, S61

Weather, power of, S88
Wedge, S93–94
Weight
    definition of, S53
    and gravitational potential
        energy, S115
    and gravity, S53–54
Wheel and axle, S100–101
Work, S84–85, S122
    definition of, S84
    effect of machines on, S89–91
    and force, S84–85
    measurement of, S85
Work input, S89–90
Work output, S90
Wright, Orville and Wilbur, S61
Watt (W), S86–87

Credits

Cover Background: Ken Karp
Photo Research: Omni-Photo Communications, Inc.
Contributing Artists: Illustrations: Jeani Brunnick/Christine Prapas, Art Representative; Warren Budd Assoc. Ltd.; Holly Jones/Cornell & McCarthy, Art Representatives; and Raymond Smith. Charts and graphs: Function Thru Form, Gerry Schrenk
Photographs: 4 Thomas Kitchin/Tom Stack & Associates; **5** top: Stephen Dalton Animals Animals/Earth Scenes; bottom: T. J. Florian/Rainbow; **6** top: Lefever/Grushow/Grant Heilman Photography; center: Index Stock Photography, Inc.; bottom: Rex Joseph; **8** left: Joanna McCarthy/Image Bank; right: Joe McDonald/Tom Stack & Associates; **9** V. Cavataio/All Sport/Woodfin Camp & Associates; **10** and **11** E. M. Bordis/Leo De Wys, Inc.; **12** top: Pedro Coll/Stock Market; bottom: J. Scott Applewhite/AP/Wide World Photos; **13** Culver Pictures, Inc.; **14** left: Paul J. Sutton/Duomo Photography, Inc.; right: J.A.L. Cooke/Oxford Scientific Films/Animals Animals/Earth Scenes; **15** left to right: J. Alex Langley/DPI; H. Armstrong Roberts, Inc.; G. Ziesler/Peter Arnold, Inc.; H. Armstrong Roberts, Inc.; **16** left: CNRI/Science Photo Library/Photo Researchers, Inc.; right: Kim Heacox/Allstock; **17** Terry G. Murphy/Animals Animals/Earth Scenes; **18** top: David Brownell/Image Bank; bottom: Herb Snitzer/Stock Boston, Inc.; **19** Stephen J. Shaluta, Jr./DPI; **20** NASA; **22** Charles Krebs/Stock Market; **23** Al Tielemans/Duomo Photography, Inc.; **24** top: Mark E. Gibson/Stock Market; bottom: Bruce Curtis/Peter Arnold, Inc.; **25** Jack Parsons/Omni-Photo Communications, Inc.; **26** top:

McDonough/Focus On Sports; bottom: Seth Goltzer/Stock Market; **27** left: Martin Rogers/Woodfin Camp & Associates; top right: Ben Rose/Image Bank; bottom right: © 1993, Estate of Harold E. Edgerton, Courtesy of Palm Press, Inc.; **28** left: Stephen Dalton/Animals Animals/Earth Scenes; right: David Madison Photography; **29** left: Stephen Dalton/Animals Animals/Earth Scenes; right: Ben Rose/Image Bank; **34** and **35** NASA; **36** top: David Madison Photography; bottom: David Muench Photography, Inc.; **37** top left: Marty Cooper/Peter Arnold, Inc.; top right: Neal Graham/Omni-Photo Communications, Inc.; bottom: E. Williamson/Picture Cube; **39** top left: Ben Rose/Image Bank; top right: Lenore Weber/Omni Photo Communications, Inc.; bottom: Nancie Battaglia/Duomo Photography, Inc.; **40** top: Wally McNamee/Woodfin Camp & Associates; bottom left: Focus On Sports; bottom right: Gary Ornsby/Duomo Photography, Inc.; **41** Clifford Hausner/Leo De Wys, Inc.; **42** top and bottom: David Madison Photography; **43** left: Jerry Wachter/Focus On Sports; right: ©, 1993 Estate of Harold E. Edgerton, courtesy of Palm Press, Inc.; **46** Tom Mangelsen/Images Of Nature; **48** top: Peticolas/Megna/Fundamental Photographs; bottom: Heinz Fischer/Image Bank; **52** and **53** NASA; **54** Gabe Palmer/Stock Market; **55** left: © Lennart Nilsson, THE INCREDIBLE MACHINE/National Geographic Society; center: Globus Brothers/Stock Market; right: © Lennart Nilsson, BEHOLD MAN/Little, Brown & Company; **59** NASA; **62** top: DPI; bottom: Elyse Lewin/Image Bank; **63** Fundamental Photographs; **64** Richard Hutchings/Photo Researchers, Inc.; **65** Paul Dance/Tony Stone World-

wide/Chicago Ltd.; **67** left: Ann Purcell/Photo Researchers, Inc.; right: F. Stuart Westmorland/Tom Stack & Associates; **68** David Madison/Duomo Photography, Inc.; **69** left: Ed Bock/Stock Market; right: Grace Davies/Omni-Photo Communications, Inc.; **71** Alan Gurney/Stock Market; **72** Zviki Eshet/Stock Market; **73** left: DPI; right: Michael Melford/Image Bank; **76** left: Craig Aurness/West Light; right: Jim Brandenburg/Woodfin Camp & Associates; **77** left: Split Second; right: Michael Ponzini/Focus On Sports; **82** David Burnett/Contact Press Images/Woodfin Camp & Associates; **83** David Burnett/Contact Press Images/Woodfin Camp & Associates; **84** top: Alon Reininger/Woodfin Camp & Associates; bottom: Randy O'Rourke/Stock Market; **85** Zig Leszczynski/Animals Animals/Earth Scenes; **86** T. J. Florian/Rainbow; **87** left: Mike Maple/Woodfin Camp & Associates; right: Grant Heilman Photography; **88** left: E. R. Degginger/Animals Animals/Earth Scenes; right: Bill Bridge/DPI; **89** left: John Lewis Stage/Image Bank; right: Michael Barnett/Science Photo Library/Photo Researchers, Inc.; **91** top: P. Wright/Superstock; bottom: Hank Morgan/Rainbow; **92** Grace Davies/Omni-Photo Communications, Inc.; **93** Ken Karp; **94** top: J. Alex Langley/DPI; bottom left and right: Ken Karp; **95** top to bottom: Wil Blanche/DPI; Wil Blanche/DPI; Dr. E. R. Degginger; Ben Mitchell/Image Bank; **96** left: Ken Karp; center: John G. Ross/Photo Researchers, Inc.; right: Lawrence Fried/Image Bank; **97** Focus On Sports; **98** Steve Hansen/Stock Boston, Inc.; **99** top: Hal McKusick/DPI; bottom: Gabe Palmer/Stock Market; **100** left: Paul Silverman/Fundamental Photographs; center: Elyse Lewin/Image Bank; right: Hugh Rogers/

Monkmeyer Press; **101** Richard Megna/Fundamental Photographs; **106** and **107** NASA; **108** left: David Woods/Stock Market; center: Hans Wendler/Image Bank; top right: Thomas Kitchin/Tom Stack & Associates; bottom right: Mitch Toll/Sygma; **109** left: Bryce Flynn/Stock Boston, Inc.; right: Jean-Paul Nacivet/Leo De Wys, Inc.; **110** top: Obremski/Image Bank; center: Randy O'Rourke/Stock Market; bottom: John Blaustein/Woodfin Camp & Associates; **111** NASA; **112** top: Wil Blanche/DPI; center: Robert E. Daemmrich/Tony Stone Worldwide/Chicago Ltd.; bottom: George Munday/Leo De Wys, Inc.; **114** top left: Kim Golding/Tony Stone Worldwide/Chicago Ltd.; top right and center: David Madison/Duomo Photography, Inc.; bottom: Gary S. Chapman/Image Bank; **115** top left: Jack Parsons/Omni-Photo Communications, Inc.; top right: John Kelly/Image Bank; bottom: David W. Hamilton/Image Bank; **116** Bruce Curtis/Peter Arnold, Inc.; **117** Comstock; **118** Thomas Kitchin/Tom Stack & Associates; **119** Ken Karp; **121** California Technical Archives/California Institute of Technology; **122** top: C. E. Miller/M.I.T./North Point Video; bottom: Peter Menzel/Stock Boston, Inc.; **123** Maurits Cornelis Escher/Art Resource; **124** left: Ray Mathis/Stock Market; right: Thomas Dimock/Stock Market; bottom: Spencer Swanger/Tom Stack & Associates; **125** left: Philip Wallick/Stock Market; right: Craig Tuttle/Stock Market; **130** and **131** NASA; **132** Tony Stone Worldwide/Chicago Ltd.; **133** Hank Morgan/Rainbow; **135** NASA; **136** David Woods/Stock Market; **157** Manfred Kage/Peter Arnold, Inc.; **159** Ben Rose/Image Bank